Elements of the Scientific Paper

Elements of the Scientific Paper

MICHAEL J. KATZ

YALE UNIVERSITY PRESS
New Haven and London

Published with assistance from the Louis Stern
Memorial Fund.

Selections from the following poems have been
reprinted by permission:

Reuel Denney, "The Laboratory Midnight." From
Connecticut River and Other Poems by Reuel
Denney. Copyright © Yale University Press, 1939.

Archibald MacLeish, "Reasons for Music." From
Songs for Eve by Archibald MacLeish. Copyright
© 1954 by Archibald MacLeish. Reprinted by
permission of Houghton Mifflin Company.

T. S. Eliot, "Little Gidding." From *Four Quartets* by
T. S. Eliot, copyright 1943 by T. S. Eliot, renewed
1971 by Esme Valerie Eliot. Reprinted by permission
of Harcourt Brace Jovanovich, Inc.

Designed by Margaret E.B. Joyner
and set in Times Roman and Clarinda Typewriter
type by Rainsford Type.
Printed in the United States of America
by Halliday Lithograph, West Hanover,
Massachusetts.

Library of Congress Cataloging-in-Publication Data
Katz, Michael J.
    Elements of the scientific paper.

    Includes index.

    1. Technical writing.   I. Title.
T11.K34   1985   808'.0666   85–40464
ISBN 0–300–03491–1 (alk. paper)
ISBN 0–300–03532–2 (pbk. : alk. paper)

The paper in this book meets the guidelines for
permanence and durability of the Committee on
Production Guidelines for Book Longevity of the
Council on Library Resources.

10   9   8   7   6   5   4   3   2   1

To my wife

Catherine Page Maxwell Katz

Contents

ACKNOWLEDGMENTS

Sidney Katz pointed out to me the fundamental importance of tables in a scientific paper. Stuart Geman and Toyoko Yamashita kindly read through the section on statistics in the chapter on numbers and contributed helpful suggestions. William Eddy read the chapter on numbers, provided a list of useful statistics texts for the appendix, and advises scientists whose work depends on statistical analyses to "consult a statistician early and often, and begin *before* you collect data." Edward Bloch and Raymond Lasek read drafts of the entire manuscript, and their comments have led to a much improved final version. In addition, I have incorporated a number of Lasek's specific ideas into the chapter on editors, referees, and revisions. I am also indebted to the Alfred P. Sloan Foundation and to the Whitehall Foundation for their support during the writing of this book.

Introduction

Writing a scientific paper is a necessary part of the scientific endeavor. Science is a communal exercise—a continuous conversation among a community of people—and scientific papers are its monologues. The essence of science is its fraternity.

Scientific knowledge is a dynamic web, maintained by the activities of scientists who are its living nodes, constantly at work resculpting and infusing it with energy. These nodes—the scientists—are linked through the permanent records of science, largely the scientific papers. By themselves, experiments, observations, and insights are ephemeral phenomena, isolated islands of data, but scientific papers preserve this information and relay it to others. Scientific papers set data in a context, and they bridge the spatial and temporal distances separating scientists.

This book offers a set of rules for writing scientific papers, and it presents these rules in a recipe that is a miniature of the scientific process itself. The writing of a paper should parallel the sequence of scientific experimentation: First, set the boundaries of your survey and delimit the realm of exploration. Second, construct the tools. Next, collect the data and translate them into a common, standard, and simple form: organize your observations into a coherent structure with its own internal logic. Now, set the results in a broader context—connect the structure that you have created with the logical structures of other scientists. Finally, polish the entire work by pruning unnecessary words and statements, smoothing rough and awkward passages, and tightening the logic. This is the process of building a scientific paper, and to illustrate these various steps specifically, I have built a short paper, step by step, within the text. I am a biologist, and the examples that I present are biological, but the principles will hold across the natural sciences.

Scientific papers are a very specialized form of writing; nonetheless, they are a form of narrative. A scientific paper cannot be a mere collage of facts and numbers; it must be readable and it must partake of the art of good writing. For this, there is no better guide than the sparkling little book *The Elements of Style* by William Strunk, Jr., and E. B. White.

The most difficult part of scientific writing is to construct a text that is at once simple and thoughtful. My advice is to write your paper, to let it gestate for a week or two, and then in a relaxed moment to ask yourself: What exactly is the point I wish to make? After the gestation, when you take the final step in creating a scientific paper, when you sit down to revise and reconstruct, aim the entire document toward the answer to this question. Simplify and clarify, and write a paper that is disarmingly straightforward and to the point. The most profound compliment you can get is the comment from a colleague, "But of course, it's all so obvious."

Elements of the Scientific Paper

1 / What is a Scientific Paper?

Science is what the world is, earth and water.
And what its seasons do. And what space fountained it.
It is forges hidden underground. It is the dawn's slow salvo.
It is in the closed retort. And it is not yet.

Reuel Denney, "The Laboratory Midnight"

A scientific paper is prose literature: like other narrative forms, a scientific paper communicates one person's ideas to another through the written word. At the same time, a scientific paper is also like a poem—it is constrained within a well-defined form, and it best succeeds when it accomplishes its aims entirely within the preestablished bounds.

Much of the scientific endeavor is natural history, and the specific things that a scientific paper communicates are particular observations. A scientist typically spends a significant part of his life observing: he might watch insects in the field, cells in the laboratory, or abstract structures in a computer. In each of these settings, the scientist plays his role as a natural historian of some class of phenomena, and the scientific literature plays its role as a vast repository of natural historical observations.

Beyond this role, beyond being a standardized form for record keeping, scientific papers serve a more fundamental, more difficult, and ultimately more important function. Scientific papers are explicit logic arguments that interweave disparate natural historical observations. Scientific papers form the scaffolding of science. In science, it is not sufficient to announce that observation A must be related to observation B. The scientist must visibly construct the connections between A and B, and those connections must be fabricated according to certain prescribed rules, that is, according to scientific logic. A scientific paper explicitly describes a segment of this scaffolding of connections. At its best, a scientific paper is a closely reasoned set of logic steps that weave an interconnection between observation A and observation B.

The constraints that dictate the form of a scientific paper all conspire to make it an effective linkage in the meshwork of synthetic science. First, the scientific paper must have a particular format, usually: abstract, introduction, materials and methods, results, discussion, conclusions, references. This format is a standard linkage form, an outline forcing each scientist to face the same basic questions and to attempt roughly comparable analyses for even the most varied situations.

Second, the language and the style must mirror the highly stylized and bland formulations of mathematical logic. Much of the color of our everyday language derives from ill-defined, emotionally charged, ear-tickling images conjured up by the nuances available in a large vocabulary of sensual and particularly human words, words such as *pie in the sky*, *piffle*, and *piggyback*. In contrast, the goal in a scientific paper is to reduce the intangible and the implicit meanings of our words. Scientific words should be the words of logic, because, when the operations of mathematical logic are used to build a scaffolding of interconnections—even quite intricate chains of interconnections—we can routinely trace through all of the paths and we can have confidence that each of the individual links is strong.

Third, a scientific paper must be published in a journal or a book. By inscribing the explicit details of their observations and arguments in widely disseminated media, scientists subject their logic to independent and critical scrutiny. The unadorned announcement that a connection exists between A and B leaves other scientists with nothing but the bald statement as a building block. On the other hand, a description of the details of the connection between A and B gives other scientists the rich and subtle inner structure of that connection on which to work as they go about their business of further remodeling the scientific edifice, and a published record of these details gives the descriptions solidity and permanence.

In summary, the substance of a scientific paper is natural historical observation, the task of a scientific paper is to detail logical connections between specific observations, and the three major constraints for a scientific paper are: format, language, and publication.

2 / Why Write a Paper?

The labor of order has no rest:
To impose on the confused, fortuitous
Flowing away of the world, Form—
Still, cool, clean, obdurate,

Lasting forever, or at least
Lasting: a precarious monument
Promising immortality, . . .

Archibald MacLeish, "Reasons for Music"

Why write a scientific paper? After performing an experiment, you have peered into new corners of the natural world, you have come to some understanding of the phenomena studied, and you now wish to communicate your insights to others. At this point, you have accumulated a pile of data, but it is still in rough form. By writing a paper, you complete the scientific exercise, building a neat and thoughtful packet of information that has its own place in the broad web of science.

Most sets of observations can form the bases for legitimate scientific papers. An experiment is a natural history exercise, an organized collecting expedition, and whenever you explore a new realm you can find something of scientific interest. Artistically, of course, you can explore and collect simply for the fun of exploring and collecting—you can walk the same route through the park or play the same piece of music time and time again just for the pleasure of the exercise. Scientifically, however, you cannot repeat the same experiment over and over again and still make a significant contribution. When the observations have already been thoroughly recorded, or when they can be completely predicted from first principles (such as the principle that 10,000 flips of a true coin will always produce about 5,000 heads), then the exercise of collecting data is not worthwhile. In all other cases, the observations are potentially interesting to scientists and can be the fabric of a legitimate paper.

Notice that, at this stage, observations need not be judged important to be worth writing about. Import is not inherent in data themselves; it is instead a quality imparted by people, and it reflects the transient fads, the trends, the quirks, and the times of a science. Scientific results can be important because they are broadly reaching—the essentially ubiquitous role of DNA made its discovery important. Scientific results can also be important because they fit into some special niche in the scientific edifice, filling an irritating gap. The actual sighting of the planet Neptune in 1846 was not broadly synthetic because its existence had been previously and accurately predicted; nonetheless, the confirmation of the prediction was important because it was scientifically satisfying. At the same time, scientific results can be important simply because scientists happen to be interested in them, not because of any special intrinsic merit. For example, scientists have always been fascinated by biological oddities, and for this reason they attach an importance to such recent findings as carnivorous mushrooms and square bacteria.

The importance of any observations is always laid upon them by the surrounding scientists. Thus, as the source of a set of new observations, you have the two tasks of putting your hard-earned data into a form that can be understood by other scientists and of forging at least the initial links between your data and other extant observations. In other words, you begin to invest observations with import by writing a good scientific paper about them.

A scientific project is not complete without a written record, and the reason for writing a scientific paper goes far beyond solid- ifying ephemeral ideas and preserving transient observations. To write a good paper, one must construct an orderly arrangement of data and frame a complete sequence of logic, from data to conclu- sions. These processes are critical parts of any complete research project. Data are useful to others only when they are neat and intelligible, and data take on meaning only in a fully defined context. Moreover, synthetic statements (inferences and conclusions)—the broad building blocks of science—must be erected carefully and neatly around complete and meaningful data.

Writing a scientific paper is tantamount to undertaking these final and requisite scientific exercises, and publishing the paper in a reviewed journal helps to ensure that the entire research project— and especially the final desk work—has been done thoroughly and in a way that other scientists can understand. Scientific writing cap- tures and substantiates a part of the natural world; this is necessary for good science, but it is a mixed blessing. Science tames the wild and the arcane in nature, but it also drains some of the magic from the world.

EXAMPLE: How Straight Do Axons Grow?

To illustrate the process of writing a scientific paper and as a forum for discussing the details of that process, I will build a small man- uscript step by step. Because of my background, my interests, and the tools at hand, I have chosen the problem: "How straight do axons grow?" As an underlying set of data, my observations will comprise records of the growth paths of axons in tissue culture dishes, with concomitant assessments of their straightness. Parts of this paper are distributed throughout the text of this book; the entire paper is included as Appendix A.

3 / Choosing the Appropriate Forum

Writing a paper is a good way to understand your data. The skeleton of a scientific paper poses an organized set of questions, and as you fill in the sections of your paper you are automatically guided through an orderly analysis. The details of this analytic skeleton differ somewhat among the various published forums; it is therefore a good idea to choose your particular forum early so that you can work from an exact guide.

SCIENTIFIC JOURNALS AND COLLATION VOLUMES

There are two main forums for scientific papers: scientific journals and collation volumes. Journals are the mainstays of scientific communication, and they are the repositories of the bulk of the scientific data. Journals are sponsored by scientific organizations, have independent editorial boards composed of scientific scholars, and are refereed forums. The papers printed in a scientific journal are approved by at least one independent scientist who certifies that the paper is readable and logical and that it meets appropriate standards in the field. There are no absolute standards—the editorial criteria

7

vary widely—but the scientific publishing community operates with the faith that its reviewers are conscientious scientists who will filter out the poorly presented manuscripts and recognize good scientific papers.

The journal review process has become a formative and constructive step in the scientific endeavor to categorize observations and build synthetic abstractions. Reviewers suggest changes in manuscripts and improvements in the experiments, and the journal editors often require that the manuscript or the experiments be revised before the paper is published. In this way, papers in scientific journals meet the scientific standards of their field; and, through the review process—through iterative interactions with independent objective peers—these papers become improved scientific statements.

The second major forum for scientific papers is the collation volume. Collation volumes are bound collections of scientific papers that have been assembled around a particular topic. The volumes can be one-time-only monographs, or they can be parts of a series, with such titles as *Advances in Experimental Medicine and Biology*, *Annual Reviews of Physiology*, *Contemporary Topics in Immunology*, *Handbook of Experimental Pharmacology*, *International Reviews of Cytology*, *Methods in Enzymology*, and *Progress in Brain Research*. As a rule, contributions to collation volumes are invited: an editor asks a group of scientists to contribute papers. Collation manuscripts are not always independently reviewed, and the initial invitations serve as the primary scientific filters.

In a collation volume, the format for a scientific paper is usually less restrictive than the format required by a scientific journal. Although collation volumes sometimes tolerate articles written in an essay style, most scientific journals adhere to a fairly standard form (usually: abstract, introduction, materials and methods, results, discussion, conclusions). I will focus on the construction of a paper that is appropriate for a typical scientific journal.

CHOOSING A SCIENTIFIC JOURNAL

The subject matter of a scientific paper dictates the appropriate circle of scientific journals, and that circle is usually quite small. Major discoveries and reports of very general interest fit into a few broad-ranging journals such as *Science*, *Nature*, and *Proceedings of the National Academy of Sciences (USA)*. On the other hand, most reports fit into specialty journals. For example, embryological observations might appear in journals devoted to developmental biology, such as *Anatomy and Embryology*, *Developmental Biology*, the *Journal of Embryology and Experimental Morphology*, or the *Journal of Experimental Zoology*. Alternatively, embryological observations can appear in journals devoted to the particular system that has been studied; if the observations were on developing nervous systems, then appropriate journals would also include *Brain Research*, the *Journal of Comparative Neurology*, the *Journal of Neuroscience*, and *Neuroscience*.

History is a good guide to the future, and the best way to discover the appropriate circle of journals is to look at a wide range of journals and to find recently published articles that are like the one you plan to write. Find articles with subjects similar to yours and with styles compatible with the experiments you have done. Each journal has an implicit style. Some journals, such as the *Journal of Neurocytology*, emphasize morphology—structural observations, pictures, descriptions of the form of things. Some journals, such as the *Journal of Neurochemistry*, emphasize compositional analysis—chemical fractionations, tables and graphs, descriptions of the elemental components of things. Others, such as the *Biophysical Journal*, emphasize quantitative analyses—mathematical characterizations, formulas, numerical descriptions of the relations between things.

After you have found a circle of journals covering appropriate subject areas, match your experiments and your data to the journals' styles and you will find that you have now narrowed your list to only two or three appropriate journals. When you make your final choice as to where to send your manuscript, an outside opinion will help; here, as at many other stages in writing your paper, consult a colleague—ask an objective observer in which journal he or she thinks your paper would best fit.

EXAMPLE: A JOURNAL FOR AXON GROWTH MEASUREMENTS

As an example, I am writing a brief paper, and for this I have chosen to report a set of detailed measurements on axons growing in tissue culture dishes. These data will then be followed by a mathematical characterization of the growth paths of the axons, assigning each path a number that quantifies how straight the axon has actually grown. (The paper that I compile in the text is an excerpted and modified version of a full article that is included in Appendix A.)

I have not made any very general discoveries, so I begin by considering specialty journals—journals of neurobiology, of cell growth, and of cell movement. My data consist of a table of measurements, a mathematical analysis, and some statistical summaries. Looking through the library, I find familiar-looking papers in *Brain Research*, in *Cell Motility*, and in the *Journal of Neuroscience*. The overall format of the relevant articles is similar in all three journals.

Asking colleagues, I find that *Brain Research* and the *Journal of Neuroscience* tend to be read by scientists whose primary interest is nervous systems. *Cell Motility* tends to be read by scientists whose primary interest is the locomotion of cells and of cell processes (such as axons). To which audience should I aim my paper? I am a neurobiologist, so I choose the *Journal of Neuroscience*.

4 / Writing with a Computer

Computers can work faster than people, can store more specific details than most human brains, and can repeat identical tasks forever without getting bored or making significant mistakes. In science, computers make calculations and enact simulations; and, today, computers as word processors aid and abet the scientific writer.

A computerized text is like a set of mosaic tiles. Words, phrases, and paragraphs can be arranged and rearranged in any sequence. New details can be grafted anywhere, and effete ideas can be scooped out and discarded. The writing can be reworked endlessly, and with a bit of care the final product will always have a neat and polished appearance.

When composing on the computer, I suggest that you build your paper in discrete blocks—make each section a separate document—and compile the full manuscript only in the very end. The major sections of a paper provide a standard outline that will organize your overall presentation. As a start, set down an outline for each separate section. On the computer, I begin with a working outline and then literally overwrite it with narrative. Because it is so easy to delete, prune, reword, and edit a computerized text, I write the first version of each section with abandon and let the early drafts be a rush and jumble of ideas. Inevitably, the cold hard light of a new day will temper the text, and because of the ease of making extensive changes and restructuring entire sections, the final manuscript eventually becomes tight and coherent.

As guides, set out working outlines at all levels; you can even outline paragraphs. In scientific papers, each paragraph should be a unit of logic, and the ideal unit of logic is the syllogism. Classic syllogisms present two observations and then draw a conclusion—for instance:

> Drug A depolymerizes microtubules.
> If axonal microtubules are depolymerized, then the axon will retract.
> Thus, I predict that drug A will cause axons to retract.

Syllogisms proceed in the natural order of discovery: first, you uncover one clue; next, you uncover a related clue; finally, you connect the two clues and build a hypothesis.

As an example, consider the second paragraph of the sample scientific paper of Appendix A. The underlying syllogism is:

> Neurons grown in homogeneous environments would reveal intrinsic configuration determinants.
> Tissue culture provides a homogeneous growth environment.
> Thus, axons in tissue culture will reveal intrinsic configuration determinants.

After some amplification and polishing, the final version became:

```
If a neuron could be grown in an ideal homogeneous environ-
ment—an environment with no extrinsic guiding contours—
then its axon should assume a configuration that was largely
determined by intrinsic factors. Tissue culture systems can
provide relatively homogeneous environments, and by growing
axons in dispersed culture on homogeneous surfaces one may
be able to reveal the intrinsic determinants of axon
configurations.
```

Use the syllogistic form to outline troublesome paragraphs. Your entire paper cannot be a string of syllogisms, but when you find that your explanations are becoming long and convoluted, when you write more than one "because," "thus," or "therefore" in a paragraph, when you are unsure about the order in which to present a complex set of observations, or when you are simply stumped as to what to say next, try breaking your argument into syllogistic paragraphs. Outline your facts and observations as a list of logical couplets, each with a direct conclusion. Then overwrite your outline of these syllogisms, making each one into a separate readable paragraph.

Computers help to overcome the inertia of rewriting and reworking—this is their major strength. Computers also do boring tasks. Many word processing programs will find all the occurrences of specific words or phrases in a text. I use this function to cleanse my manuscript of words that I habitually misuse. For example, I usually search my final documents for the word *data*, which I have invariably followed by a singular verb instead of the correct plural verb. The chapter on language contains a list of commonly misused words, and you may wish to check your texts regularly for those errors that occasionally slip into your writing.

Writing is research; it is a process of discovery. Although you begin with a tentative outline, each scientific paper has a natural, logical structure that emanates from within and cannot be completely imposed from without. The particular set of results at hand will fit best only in certain natural forms. These forms are flexible but not infinitely malleable, and you cannot arbitrarily construct your final discussion or your introduction according to preconceived notions. It is unavoidable that you will bring presumptions, biases, and hypotheses to your early drafts, but forging a coherent story from the real data can bend even strongly held beliefs, and a natural internal logic will slowly emerge. Reworking and reexamining the detailed arguments of your paper will bring out its inherent form, and continued rewriting will make the text scientifically cogent and true.

This is where computers shine. Computerized text is a dynamic, living medium; the sentences grow and shrink and metamorphose, and the computer puts energy into the document, continuously neatening, reshaping, and cleaning the text. At each stage you have a finished product; you can start your innumerable rewritings with a fresh copy, and you can work undistracted by smears and smudges, cross-outs and appended insertions, or scraps of paper cut and pasted here and there, dotting your written landscape like golf flags. Computers encourage the most important of all writing practices—rewriting.

5 / Language

"That's a great deal to make one word mean," Alice said in a thoughtful tone.

"When I make a word do a lot of work like that," said Humpty Dumpty, "I always pay it extra."

Lewis Carroll,
Through the Looking-Glass and What Alice Found There

A scientific paper is prose—it must be a readable narrative—and, as a practical guide and as an inspiration for narrative writing, there is no better essay than Strunk and White's *Elements of Style*. At the same time, a scientific paper is a special form of narrative writing; it is not literary prose. Instead, it has special constraints and is built with its own peculiar language. The primary purpose of a scientific paper is not to speak to the heart but to the brain.

In science, a smooth, flowing style is helpful, and balanced and cogent wording is an aid; but the essence of scientific style is crystal clarity. Each sentence must convey a definite idea, and it must have an unequivocal interpretation: there can be no mystery, no vagary, and no intimations of unwritten meanings or of arcane knowledge. In science, the medium is not the message. The medium is a standardized format for presenting new data in a fashion that will most directly and immediately fit into the broader, fairly stereotyped schema of other scientific data.

For clarity, write in short sentences and pare down all excess phrases. "The fact that axons have been observed to grow randomly suggests" should be trimmed. "The random growth of axons suggests" is better. Write: "it showed" not "it served to show." "The design of the present study allows for the determination of whether or not an axon has random growth" reduces to "This study can identify the random growth of an axon." "The theory has been subjected to modification" should be simply "The theory has been modified." "provided support for" should be "supported." Change "subjected to radiation" to "irradiated." "It was found that the cells" or "It was demonstrated that the cells" is best written as "The cells." "Cell staining was performed" should be "cells were stained"; and "in a linear fashion" is simply "linearly."

Write in the active voice when possible. "The cells protruded pseudopods" is better than "Pseudopods were protruded by the cells." Also, do not be afraid of using first-person singular pronouns when they are appropriate. "I propose" (or "we propose") is better than "it is proposed." For single author papers, do not use "we" or "our," unless you are actually referring to things shared by others.

WORDS

Use simple, direct words, words with little emotional weight and clear meanings. Science is already bursting with new terminology, and new words and hyphenated neologisms only clutter an article and confuse the reader. "Interrater reliability" is a poor substitute for "reliability between raters." Write: "animals exposed to ethanol," not "ethanol-exposed animals."

Contemporary science is written in English; use English forms whenever possible, not Latin or Greek. Latin names are still the standard for taxonomy, but English forms (such as *chordates* for *Chordata*) can sometimes be substituted. Today, English is more

immediately and more widely understood than is Latin or Greek—
for example, *midbrain* is a more useful term than *mesencephalon*.

Ad hoc

Ad hoc means "for this special purpose." An ad hoc committee is
a committee specially convened for a particular purpose, and ad hoc
assumptions are assumptions chosen expressly for the situation at
hand. *Ad hoc* is not necessarily synonymous with *temporary, casual,*
or *without substantial basis.*

Alternate. Alternative

As adjectives, *alternate* means every other member of a series and
alternative means another mutually exclusive possibility. "We chose
the alternative explanation," not "we chose the alternate
explanation."

Arbitrary

In the sciences, *arbitrary* describes an action carried out by whim,
without plan or organizational design. If you choose experimental
subjects by blindly pulling them from a pool, you choose them
arbitrarily, not randomly. Random actions operate under a specific
design. (See **Random** below.)

Compose. Comprise. Constitute

Comprise means to contain. The alphabet comprises letters. In contrast, letters constitute or compose an alphabet.

Constant. Continual. Continuous

Constant means "steady and unceasing." *Continual* describes a frag-
mented stream of events (discrete events) and means "repeating
over and over again." *Continuous* describes an unbroken stream of
events and means "uninterrupted and without pause."

Correlate

Correlate ("relate together") is an umbrella term comprising innumerable different types of linkages. If possible, be more specific. Write: "the time course of action of drug A overlapped that of drug B" or "the cellular changes caused by drug A appear identical to those caused by drug B" or "drug A affected membrane constituents at the same locations as drug B" or "drug A and drug B each produced the same toxicity syndrome." Do not simply state: "the effects of drug A correlated with those of drug B."

Data

The word *data* is a plural: "Data were collected," not "Data was collected." The singular *datum* is awkward; use *observation*, *result* or, better yet, be specific ("the length," "the time interval," "the voltage," or "the velocity").

Deduce. Induce. Infer

To *deduce* is to particularize; to *induce* is to generalize; to infer is to conclude by logical reasoning. Deduction and induction are both types of inference. Use *infer* and *inference* unless you wish to distinguish one particular type of logical reasoning.

Definite. Definitive

Definite means "precise, clear, and well-defined"; *definitive* means "final, complete, and substantial." "Definite results" are exact and unquestionable; "definitive results" will never be surpassed.

De novo

De novo is often used to mean "totally without precedent," but it actually means only "afresh," "again," "anew," and "from the beginning." For precise writing, it is best to avoid *de novo*.

Employ. Use. Utilize

Scientific papers need a minimum of color and variety. Use simple words like *use* whenever possible. *Employ* means "engage the services of" or "hire." *Utilize* has the connotation of making a practical use of something.

Enhance. Increase

Enhance means to increase the contrast; *increase* means to enlarge the value. "The volume was increased," not "the volume was enhanced."

Hypothetical. Putative

Hypothetical is the less restrictive of these two terms; it means "conjectured on the basis of some specified model or hypothesis," regardless of the acceptability of the hypothesis. *Putative* means "suggested candidate for," and it implies a substantive basis for the suggestion. A hypothetical precursor is any precursor that we can imagine and define, whereas a putative precursor is a precursor that we also consider to be likely.

Implicate

As conventionally used in science, *implicate* means "suggest involvement." When using *implicate*, specify the exact involvement. "These results implicate neurotrophic factors in limb regeneration" is better than "these results implicate neurotrophic factors," but "these results implicate neurotrophic factors as regulators of the rate of limb regeneration" is best.

Imply. Indicate. Intimate. Suggest

In the sciences, conclusions are frequently softened with words such as *imply*, *indicate*, *intimate*, and *suggest*. (Related qualifiers include *appear* and *seem*.) When appropriate, use *suggest* or *indicate* rather than *imply*, which has the connotation of something hidden and unwritten. *Intimate* is a literary word, and it never fits comfortably in a scientific paper.

On the other hand, stronger conclusions are better. If it is necessary to qualify the conclusion, then try to be more specific. "Axon growth paths with fractal dimensions between 1.00 and 1.31 suggest nonrandom growth" is better written as "Axon growth paths with fractal dimensions between 1.00 and 1.31 have less than a 5% probability of being random walks."

Important. Interesting

When you qualify something as important or as interesting in a scientific paper, you owe the reader an explanation. By themselves, *important* and *interesting* are too general; always follow *important* or *interesting* with "because . . . "

Inherent. Innate. Intrinsic. Endogenous

Inherent, *innate*, and *intrinsic* refer to properties that are built into and characteristic of something. *Inherent* connotes a permanent property—"inherent elasticity." *Innate* is used only for living things—"innate logic of the mind," but "inherent logic of a circuit." *Intrinsic* connotes a variable property—"intrinsic metabolic rate" or "intrinsic worth." *Endogenous* means "originating from within."

Locus

Locus has specific technical meanings: it is the position of a gene on a chromosome, or it is a set of points satisfying an equation. In more general contexts, use *locality*, *location*, *place*, *point*, or *position*.

Necessary. Sufficient

Necessary means "absolutely required for a given result"; "drug A is necessary for full recovery" means that full recovery will not occur without drug A. *Sufficient* is a less extreme condition and means only "will produce a given result"; "drug A is sufficient for full recovery" means that full recovery may occur under other conditions as well. On the other hand, *sufficient* also implies that you have a complete roster of causes. Thus, "drug A is sufficient for full recovery" means that drug A by itself will produce full recovery.

Normal. Standard. Usual

Be as specific as possible. "The most frequently observed" is better than "the standard" or "the usual." Remember, *normal* refers to a very specific distribution of numerical values—the smooth, bell-shaped curve of an equation of the form $y = Kexp(-x^2/2)$.

Obviously. Of course. Certainly. Indeed. In fact

Avoid judgmental and chatty asides.

Parameter. Variable

Parameters are any well-defined features of a system that can be varied. *Variables* are quantitative features that can take on different numerical values. Temperature can be a parameter and a variable; texture can only be a parameter.

Random

Random has a specific meaning; do not use it loosely. In the materials and methods section of a scientific paper, when your experimental design includes randomized items, you must use an objective source of random numbers—a table of random numbers (see Appendix H) or a pseudorandom number generator—to assign values to the items. If you use some other method to assign values to the items, then state the method explicitly. Write: "Patients were assigned alternately to one of the two study groups, in the order in which they presented themselves to the clinic." Do not write: "Patients were randomly assigned to one of the two study groups."

In the results, describe complex observations in detail and do not use *random* as a crutch. Complex patterns are heterogeneous or disorganized in particular ways. In the narrative descriptions, tell us exactly what these ways are for your system. Why do the observations appear chaotic? Are the nerve cells scattered among the others in an unpredictable pattern, or are the cells found in clumps of all possible sizes?

Whenever you are characterizing numerical values, try to match them to a pattern. For example, if the experimental results are normally distributed, then say so. If you cannot find a pattern that matches your observations, then that is all you are justified in saying—you cannot call your results *random*. Remember, *random* is itself a particular pattern, and to describe experimental values as random means that they fulfill specific statistical criteria of heterogeneity.

Sensitivity. Specificity

A *sensitive* test is a test that recognizes most marked individuals; a *specific* test is one that recognizes most unmarked individuals. A sensitive test may mistakenly recognize unmarked individuals, but it will ensure that you capture the vast majority of marked individuals. A specific test may miss marked individuals, but it will rarely make a mistake when it does classify an individual as marked. Ideal tests are both sensitive and specific.

Significant. Significantly

Reserve these qualifiers for "statistically significant," and use them only when the appropriate statistical tests have been applied. In other settings, be specific. Do not say only that the results differed *significantly*; tell us whether they differed by two-fold or by one-hundred-fold. Do not say that the color was *significantly* different; tell us whether the items changed from green to blue-green or from green to purple.

Since

In scientific papers, avoid ambiguities. Use *since* as an adverb or a preposition referring to time: "since beginning the experiment" or "the data have since been modified. Use *because* as the conjunction that introduces a reason: "the numbers were squared because," not "the numbers were squared since."

That. Which

That is used when you are separating out certain elements from a larger set of possibilities. "The drugs that caused an adverse reaction had bacterial contaminants." This statement explicitly limits the drugs being discussed to that subset causing adverse reactions. In contrast, *which* does not further partition the set of items under consideration; rather, it adds additional information about all of them. Usually, *which* begins a parenthetical clause that can be set off with commas from the main sentence. "The drugs, which caused adverse reactions, had bacterial contaminants." This statement means that *all* of the drugs being considered caused adverse reactions.

Valid

Try to avoid *valid*, *validity*, and *validate*; often they are weak words and can be replaced by descriptors that are more instructive and more specific to the particular situation at hand. "Seventy-five percent of the predictions were accurate, and this success rate is consistent with the hypothesis that" is better than "Seventy-five percent of the predictions were accurate, and this success rate validates the hypothesis that." On the other hand, when you mean something stronger, like *prove*, then write *prove*, not *validate*.

Yield

As a verb, *yield* has the connotations of surrendering and reluctantly relinquishing. *Give* or *produce* are usually better in scientific papers.

Appendix I lists thirty of the most common nontechnical scientific words.

6 / Numbers

> If we search the examination papers in physics . . . for the more
> intelligible questions we may come across one beginning some-
> thing like this: "An elephant slides down a grassy hillside . . ."
> . . . yet from the point of view of exact science the thing that
> really did descend the hill can only be described as a bundle of
> pointer readings. (It should be remembered that the hill also has
> been replaced by pointer readings, and the sliding down is no
> longer an active adventure but a functional relation of space and
> time measures.)
>
> A. S. Eddington, *The Nature of the Physical World*

Science takes the fuzziness, the mystery, and the poetry from the
world by using crisp and colorless language, and the crispest and
most colorless language is mathematics. A scientist attempts to capture his world in a stereotyped standardized idiom, an idiom that
depends as little as possible upon the variations and the whims of
the observers, and numbers offer an ideal language. There is little
doubt that when Galileo reported four moons around Jupiter, he
meant four, not three or five.

These numbers, the numbers that fill scientific papers, all represent pointer readings: they are tied to the real world as values

read from actual or proposed scales—pH meters, balances, watches, photometers, volt meters, scintillation counters, and rulers. Numbers, particular pointer readings, take their power from two properties: they can be ordered, and they can be embedded in a continuum. The former property allows us to unequivocally make the three elemental comparisons: greater than, equal to, and less than. The latter property allows us to generalize by infinite tiny automated steps, producing the tremendous range and capacity of mathematical induction.

For these reasons, quantify your observations.

ALGORITHMS

Numbers immediately lend themselves to the clearest of descriptive forms, the algorithm. Algorithms are machine recipes—computer programs—describing exactly how to get from point A to point B. I suggest that you write in algorithms whenever it is possible. When describing how you obtained a particular result, list the complete sequence of steps that a machine would (in theory) follow in order to duplicate your observation, and, when presenting your analyses, explicitly enumerate all of the logic steps that an ideal computer would execute when proceeding from your observations to your conclusion.

The algorithmic form is a model for any section of your paper, but algorithms are required for the numerical analyses. When numbers appear, they should be the offspring of clearly described mechanized processes. Either numbers are the pointer readings of physical devices, which you list or describe in detail, or they are the outputs of equations, which are written into the paper or are specifically referenced. Write: "The location of the axon tip was measured with an error of ± 1 grid square on a 21 cm (diagonal length) plastic grid of 5 squares / cm," not "The location of the axon tip was accurately measured." And, write: "t-tests of the means of the logs of the fractal dimensions demonstrated that neither type of axon grew randomly ($P < 0.01$)," not "statistical tests demonstrated that neither type of axon grew randomly."

STATISTICS

In the end, numbers are always compared in small groups, singly, or in twos or threes, and, when data pour out in a waterfall of values, you must reduce the volume to a few characteristic numbers. These characteristic summary numbers are statistics.

Statistics are not the actual data; in reducing the volume of numbers, you necessarily lose some information, and the deeper you become enmeshed in statistical analyses the farther you stray from the real world. Stay close to the observations; use few and simple statistics.

Some basic statistics that are almost always relevant are:

1. the total number, N, of values
2. the range of values (report the largest and the smallest values)
3. the mean value (the sum of all of the values divided by N)
4. the modal value (the most common value)

(Note that there may be more than one modal value and that the mode can differ from the mean.)

Using these statistics, a typical summary would be: "The mean number of spots was 15.3 ($N = 100$, range $= 11$ to 18, mode $= 15$)."

Besides these basic statistics, try to describe the shape of the distribution of your numbers. The full details of any distribution shape can be hard to summarize in simple words; a picture is usually the best way to capture and to transmit a complex shape. Draw a graph—specifically, a histogram—of your numbers, plotting values on the x-axis versus the number of occurrences of each value on the y-axis. The eye is an excellent pattern detector; and, in a histogram, you and your audience can immediately see how the data are spread and clustered, whether the distribution of numbers is smooth or bumpy, and where special patterns lurk amid the mass of values. Pictures give this information directly, even for data that cannot be abbreviated by any but the most intricate, arcane, and convoluted statistics. (See also chapter 10.)

Standard Deviation

Another very common and useful statistic is the standard deviation, which is a summary of the dispersal or spread of the data. The standard deviation is the square root of the variance, and it is calculated from the sum of the squares of the differences between individual values and the mean:

$$SD = ((\sum_{i}^{N}(x_i - m)^2)/(N - 1))^{1/2}$$

where: m = mean value, N = total number of observed values, SD = standard deviation, and $x_i = i^{th}$ observed value.

When the data can be approximated by the smooth, unimodal, bell-shaped normal (or Gaussian) distribution, then 68.3% of the observed values will fall within one standard deviation of the mean value, 95.5% will fall within two standard deviations of the mean, and 99.7% will fall within three standard deviations of the mean. A typical summary statement including a standard deviation statistic would be: "The mean number of spots was 15.3 ($N = 100$, range = 11 to 18, mode = 15, std. dev. = 1.31)." The implication here is that, *if the data are normally distributed*, then about 70% of the observed cases have either 14, 15, or 16 spots.

The standard deviation is sometimes concatenated with the mean value as: "15.3 $\pm$ 1.31 spots ($N = 100$)." The usual convention is to report the standard deviation to one decimal place more than the mean. When using this form always include N. The same form is also used for presenting the mean with its *standard error*, which provides an estimate of the accuracy of the mean rather than an estimate of the spread of the entire sample. To avoid confusion, you should state which form is being used—for example, "Each value is expressed as the mean plus-or-minus one standard deviation," or "Each value is expressed as the mean plus-or-minus one standard error."

Significance Tests

Significance tests summarize comparisons of your data with other distributions (populations of numbers)—either distributions of other data or ideal distributions. Two common significance tests are the *t*-test and the chi-square test.

t-**Test.** When you can assume that your data are normally distributed, you can use the *t*-test to make significance estimates. The chances that your particular observations are a sample from a normal population with a mean of *M* can be estimated using the *t* statistic, computed as:

$$t = \sqrt{N}\ (\bar{x} - M)/SD$$

where: *M* = presumed mean, *N* = number of observed values, *SD* = standard deviation of the observed values, and $\bar{x}$ = the mean of the *N* observed values. (This is the "one-sample *t*-test.") For example, suppose that you have surveyed a large number of Ten-spotted Newts (*N* = 100) and have found that the mean number of spots on each animal is 15.3 with a standard deviation of 1.31 spots. If Ten-spotted Newts are presumed to have ten spots, then *M* = 10 and the *t* statistic would be:

$$t = 10\ (15.3 - 10)/1.31 = 40.5$$

Now, turn to a table of critical *t* values (also called "Student's *t* distribution"), and find the row that lists values for the appropriate number of degrees of freedom (*N* − 1). (A *t* table is reprinted in Appendix F.) Identify the column with the largest *t* value that does not exceed the *t* statistic you have computed. The head of that column is the significance of your observed values, assuming they have been drawn from a normal population with mean *M*. For the newt example, *P* < 0.01.

Are your differences significant? There is no magic number, no absolute *P* value, for "true significance." It is customary to report the result as "significant" if the *P* value is less than 5% (*P* < 0.05), but this is merely a standard convention. In actuality, you are assessing how well the data match a fairly restrictive model which underlies your significance test. When the *P* value is small, then your model is not a very good match for the data, and by convention "small" is considered to be *P* < 0.05 and "very small" is considered to be *P* < 0.01. In the case of the *t*-test, one important building block of the underlying model is that your sample is from a population with a particular mean M, and when the P value is small (as it is in our Ten-spotted Newt example) you can conclude that it is unlikely that the population actually has this presumed mean.

Chi-square Test. The chi-square test assesses how well your observations fit a particular frequency distribution. This distribution can be simple—perhaps it is that thing A occurs 50% of the time while thing B occurs the remaining 50% of the time—or the frequency distribution can be very complex. In any case, the chi-square statistic allows you to test the hypothesis that the actual frequencies of appearance of certain things are just chance variants of some proposed set of frequencies.

The chi-square statistic, χ^2, is:

$$\chi^2 = \sum_{i}^{N} (x_i - E_i)^2/E_i$$

where: E_i = expected number of occurrences of the ith thing (the given frequency distribution), N = number of different things to be compared, x_i = actual number of occurrences of the ith thing (the observed frequency distribution). For example, if you flip a coin 100 times and if heads occurs 37 times and tails occurs 63 times, then the chi-square statistic for testing whether the coin is fair is:

$$\chi^2 = (37 - 50)^2/50 + (63 - 50)^2/50 = 6.76$$

N (the number of types of things—that is, heads and tails) = 2, and there is one degree of freedom.

Now, turn to a table of critical χ^2 values, and find the row that lists values for the appropriate number of degrees of freedom ($N-1$). (A χ^2 table is reprinted in Appendix G.) Identify the column with the largest χ^2 value that does not exceed the χ^2 you have computed. The head of that column is the significance of the observed frequencies assuming that they are chance variants of the expected frequencies. Again, the convention is to report the result as significant if the chances are less than 5% ($P < 0.05$). For the coin-flipping example, $P < 0.01$.

The *t*-test, the chi-square test, and other significance tests come with a host of attendant assumptions, and, when using significance tests, first read a statistics book (see Appendix B), or, better yet, consult a statistician.

Statistics can objectively summarize vast sets of numbers; they can automate—turn into an algorithm—comparative statements like "x is much more than y" or "x is essentially the same as y," and this is their appropriate use. Statistics can also distort the data, and this usually happens when the statistical tests are too complex and depend for their proper interpretation on an implicit model that is built on many underlying assumptions. Most data can hold only modest statistical analyses. Beware the seduction of statistics for the sake of statistics.

7 / Materials and Methods

A scientific paper is written from the foundations up, and the materials and methods are its fundamental supports. In the center of the article—the results section—a scientific paper reports natural historical observations, but all such observations take on meaning only in the context of a well-defined system, and the materials and methods section sets the precise bounds for that system. The materials and methods are considered so basic and so important that it is the one section that reviewers never ask to be trimmed. It is a good policy to write a materials and methods section early, while you are still in the midst of your experiments. Then, the many small technical details—your tricks of the trade—are still fresh in your mind and you can record them accurately.

The materials and methods describe just that: the materials, substances, supplies, tools, instruments, appliances, and contrivances; and the methods, techniques, procedures, recipes, formulas, transactions, and algorithms. They are the stuff and the ways of science, and for this section your goal should be to write a description that is detailed and complete enough for any researcher to follow your directions and to repeat your observations successfully.

Write the materials and methods section in concise formulaic sentences, without colorful language, without extraneous comments or asides, and without indeterminate statements. As throughout your paper, avoid jargon and define specialized terms and abbreviations when they first appear. The materials and methods should be as close to an algorithm (a computer program) as possible; so work in an absolutely lean and spare style, and break the overall section into many clearly labeled parts that fit into a straightforward outline. Along the way, take a black-and-white stand on details. If other researchers are to repeat your work precisely, you cannot write "maybe" or "sometimes" or "often." State definitely that either something was done or it was not done, and when you must use a qualifier, try to quantify it: Was the procedure done 50% of the time or was it done 90% of the time? Moreover, include all of the small procedural details. If there is a doubt as to whether to include some information, err on the conservative side and add the extra facts.

EXAMPLE

The paper that I am constructing explores the straightness of the growth of axons in tissue culture dishes, and I begin my writing with the materials and methods section. First, I lay down an outline:

```
MATERIALS AND METHODS

A. Cell Culture Conditions

B. Data Collection
1. Criteria for Choosing Individual Axons
2. Measurement Procedures
3. Precision of Measurements

C. Special Computations: The Straightness of Growth
```

Next, I fill in each section methodically and thoroughly:

MATERIALS AND METHODS

A. Cell Culture Conditions

    Xenopus laevis embryos were obtained from matings of
adult frogs

[Give the exact scientific name of all organisms.]

--matings were induced by injection of human chorionic gonadotropin (HCG, Sigma Chemical Co., St. Louis MO).

[HCG is a standard abbreviation for this compound. Give the source
for any specialized substance or tool.]

Twenty-four to forty-eight hr old amphibian primary tissue
cultures

[See Appendix C for a list of standard scientific abbreviations.]

were grown by disaggregating neural tube cells of tailbud
stage Xenopus embryos and by plating these cells on acid-
rinsed glass coverslips set in the bottoms of 35 mm sterile
FALCON petri dishes (Spitzer and Lamborghini, 1976).

[Include references to sources of nonstandard techniques and to
sources with further technical details.]

Stage 28-30 (Nieuwkoop and Faber, 1967) Xenopus embryos were
removed from their vitelline membranes and were washed
through four changes of sterile Steinberg solution, pH 7.5
(Hamburger, 1960). Next, the ectoderms were peeled off and
the brain primordia were dissected free and were washed in
fresh Steinberg solution. Brain primordia were then disaggregated by soaking them for 10 min in a calcium and magne-
sium free medium (59 mM NaCl, 0.7 mM KCl, 0.4 mM EDTA, pH
7.5).

[Include complete recipes for all special solutions. Room temperature is usually assumed for experimental conditions and solutions unless otherwise specified.]

Cell aggregates were broken apart by pipetting them briefly at 5 min intervals with a fire-polished Pasteur pipette. Cell suspensions were then plated in fresh Steinberg solution supplemented with 2% fetal bovine serum, 5-10 μM nerve growth factor (NGF), and antibiotics (50 units/ml penicillin, 0.05 mgm/ml streptomycin—Gibco Labs, Grand Island NY). Approximately 1-1.5 brain primordia were plated per 35 mm culture dish, and the cells were then grown at 21°C.

[Include concentrations, times, and temperatures (when appropriate) for all experimental paradigms.]

Under these conditions, 24 hr cultures averaged 170 differentiated cells per plate and approximately ¼ of these cells were neurons.

B. Data Collection

1. Criteria for Choosing Individual Axons
     Axons selected for study were only from clearly healthy cultures and were in areas of minimal debris. The axons were all at least 100 μm long and were separated from other cells and cell processes by at least 100 μm.

[Quantify whenever possible.]

Data were analyzed only when there was "normal" independent growth—i.e., growth without major retractions and without contact with other cells.

[Define any conditions idiosyncratic to the experimental paradigm.]

The final time point in each data analysis was always taken

before or during a growth period in order to be certain that
any earlier retractions were not part of a general dying
back of the axon.

2. Measurement Procedures

   Axon growth was monitored at 400x magnification on an
Olympus IMT inverted phase microscope to which was attached
an RCA TC2000 Newvicon video camera equipped with an Olympus
FK 3.3x projection eyepiece.

[Attempt to describe the apparatus so that others can picture it.]

Time lapse video records were made on SONY BR video cassettes (¾" tape) using an NEC video recorder (VC-9507) running at ¹⁄₆₄th real time. Real time was indicated with a VICOM
V240 date/time generator.

[The materials and methods section is formulaic: it should be a direct
narrative list of items and events.]

Data were collected by stopping (freezing) the tape at an
exact time point and then measuring the location of the axon
tip. The location of the tip of the axon was measured on a
transparent grid that overlaid the video screen (PANASONIC
WV-5300). The standard units of measurement were marked on a
flexible plastic template which was placed on the screen in
order to accurately measure distances along the curved axons. The screen had a diagonal length of approx. 21 cm, the
grid was divided into 5 squares/cm, and in our system this
resulted in a scale of 2.56 µm/square for the final video
image.

[Always use metric units.]

3. Precision Of Measurements

   A static image (a calibrated stage micrometer) was re-

corded for two hours, rotated 90°, and then recorded for an
additional two hours. The image did not drift, and distances
remained undistorted by the rotation. Locations could be
read from the screen to an accuracy of ⅓ grid square (approx. 1 μm). By repeating readings for 80 separate time
points on data tapes of growing axons, however, the overall
average error was found to be closer to 2 μm.

[Give an indication of the accuracy and precision of all
measurements.]

C. Special Computations: The Straightness of Growth
    As a standardized measure of the straightness of the
growth path of an axon, an estimate of the fractal dimension
was used, based on Mandelbrot (1977 1983). (See the Appendix.) The measure D is the fractional dimension of a curve
in a plane--here, it is the fractional dimension of the path
of an axon on the planar tissue culture dish. Specifically:

$$D = \log(L/a)/\log(d/a)$$

[Although letters used as mathematical terms are generally set in
italic type in publications, they need not be underlined in your
typescript—the journal's copyeditor will mark them or give the
compositor general instructions, depending on the journal's
preferences.]

where: D is the fractal dimension, L is the total length of
the path of the axon (the sum of the distances traveled in
all of the 10 min intervals), d is the greatest distance between any two points on the growth path, and a is the average
length of a step or an observed interval of growth (a = L/n
for n time intervals). In this way, a completely straight
growth path gives a fractal dimension D = 1, and a completely random walk gives a fractal dimension D-->2. Further

details, including a computer program for all calculations,
can be found in: Katz and George (1985).

[For techniques that have been described in full detail elsewhere,
cite only widely accessible sources.]

8 / Organizing the Raw Data

Papers and sketches fill cardboard boxes along your windowsill; notebooks with measurements lean at odd angles on your bookshelves; thickets of chart-recording tapes spill from your desk drawers. You are finished observing; it is time to write, and you are ready to face your mass of data and to wrestle the raw information into presentable form as scientific results.

The next step—the transformation of raw data into a finished paper—is as great as that from proposal to completed experiment. Having accumulated a set of observations, you are only halfway through the scientific exercise, and this next part of the research—the desk work—begins when you shut the door, spread your raw data before you, and organize a coherent set of results. How should you begin?

LISTEN TO YOUR DATA

Sometimes, data fall into a simple and beautiful pattern. Kepler, facing his detailed records of the movement of Mars, discovered that by considering planets to revolve about the sun in elliptical (as opposed to circular) orbits they followed the simple rule of sweeping out equal areas in equal time periods. More often, however, real data do not immediately fit such elegant models, and the actual observations are messy and, at first glance, inconclusive.

When confronted with a tangle of results, your first goal should be to construct some natural order for them—to classify and to categorize—and not to immediately prove a particular hypothesis or to discover an underlying law. Begin by laying out the data and then creating a simple arrangement among the bits and pieces of the world that you have collected. Array your results in a pattern.

One of the easiest patterns for humans to understand and to manipulate is a line. As a first step, try to put your data in a single row. If you have compiled numbers, then search for a simple transformation by which they can be laid out along a one-dimensional spectrum. If you have amassed pictures, hunt for features that allow you to line up the figures in a row where the changing features follow in an orderly sequence. Try temporal sequences—arrange things in their natural order of occurrence. Try spatial sequences— line things up in accord with their natural position in the world. Try sequences of magnitude—order things by size.

In most cases, all of your observations will not fit neatly into one simple pattern, and you will probably discover a number of arrangements that can accommodate much of the data. Choose a pattern that embraces the widest range of observations with the least extrinsic scaffolding.

At this stage, the overriding goal is to uncover some encompassing and fairly natural order among your results. That there will be such a natural order is not a preordained fact, it is only a useful working hypothesis and it is an article of faith. As Whitehead wrote: "there can be no living science unless there is a widespread instinctive conviction in the existence of an *Order of Things*, and in particular, of an *Order of Nature*" (Alfred North Whitehead, *Science and the Modern World*). When proceeding from raw data to scientific results, this order must emerge from the data. Listen to your data, not to your own presumptions or your colleagues' theories.

FILL IN THE GAPS

With your best organizational scheme in hand, neatly set out all of your data. Are there holes in the resulting pattern? Are there irregular intervals between numbers or notable spaces between pictures? Are there experimental conditions without complete documentation? If so, consider returning to the experimentation mode and collecting the missing observations.

Research has a ratcheted feel to it, and once you have finished experimenting it is natural to be reluctant to step backward and reopen your investigations. Conspicuous gaps in the simple orderly arrangement of your data, however, should stimulate you to overcome this inherent inertia and to reactivate your investigation.

Next, look along the edges of your pattern of data. Perhaps you can extend the frontiers. Will one or two more observations complete a corner of the pattern, fill in a bay, or even build a new peninsula? If more investigative work will flesh out your pattern of observations, get up from your desk and become a natural historian once more.

SET ASIDE EXTRANEOUS OBSERVATIONS

After filling in the gaps, you can take one additional step in cleaning up your pattern of observations. Some of your data will not fit into the simple arrangement that you have constructed. Occasionally, certain values are too extreme to be put on your spectrum of results. More often, however, you will have collected a few extraneous facts—observations of a different type or under different conditions from the mainstream of your results. When you neatly line up your data, you may find a small pocket of tantalizing but incomplete results that have no direct connection with the remaining mass of observations.

In these cases, set the extraneous data aside. Carefully label the information, include a description of how it was collected, and then store it away in a file drawer. Each scientific paper should focus on a single issue and should be built around one coherent set of observations. If your extraneous data are sufficiently interesting, then they deserve to be thoroughly explored in their own special paper.

In the end, you should have a simple arrangement for most of your observations. This, the pattern that organizes your raw data, is constructed without regard to interpretation. The meaning of your results is not yet at issue, and at this stage you should work with but one hypothesis: the world is orderly. This hypothesis—Whitehead's first principle—dictates only that you begin your scientific analysis with a collection of data that can be compiled into a single coherent and orderly pattern.

EXAMPLE: FRACTAL DIMENSIONS OF AXONS

In studying the growth of axons, I have collected fractal dimensions characterizing the shapes of the growth paths of twenty different frog axons grown at room temperature (21° C). These fractal dimensions are: 1.15, 1.57, 1.01, 1.46, 1.46, 1.12, 1.13, 1.04, 1.07, 1.13, 1.19, 1.26, 1.34, 1.39, 1.14, 1.18, 1.43, 1.40, 1.08, 1.84, 1.06, 1.05, 1.18. (The values are listed in the order in which they were collected.) In addition, I have calculated the fractal dimensions of two frog axons grown at a higher temperature (26° C), and these fractal dimensions are: 1.50, 1.61.

My first task is to arrange this data in a simple (and, preferably, linear) pattern. For a set of numbers, the most obvious one-dimensional pattern is usually numerical order, and the numerically ordered set of fractal dimensions is: 1.01, 1.04, 1.05, 1.06, 1.07, 1.08, 1.12, 1.13, 1.13, 1.14, 1.15, 1.18, 1.18, 1.19, 1.26, 1.34, 1.39, 1.40, 1.43, 1.46, 1.46, 1.50*, 1.57, 1.61*, 1.84. (High temperature values are marked with an asterisk.) Now, I look for irregular gaps. Being lognormally distributed, the fractal dimensions become more widely spaced as they increase, but in general they appear to be spaced fairly evenly. Were I to have found a few particularly strange values, I would have to seriously consider generating more data in order to convince myself that the unusual numbers were merely rare occurrences.

Next, I ask whether I can extend the edges of the pattern. In this case, the question is whether the basic data set can be more complete. Sometimes, you can use intuitive methods for assessing the completeness of your results. On the other hand, statistics can be a very helpful objective guide for determining whether more data are necessary. If you are dealing in numerical data, the best plan is to talk with a statistician. In the case of fractal dimensions, a statistical consultation has suggested that at least ten examples (of growth paths containing eight or more points) are required for a complete data set. Thus, I have sufficient data to begin writing my results.

Finally, I identify extraneous observations. The high temperature results form too small a set to be anything but anecdotal results. High temperature is an additional variable, and temperature perturbations represent a new experimental paradigm. Therefore, I set the 26° C data aside for future work.

9 / Results

If the art of the detective began and ended in reasoning from an armchair, my brother would be the greatest criminal agent that ever lived. But he has no ambition and no energy. He will not even go out of his way to verify his own solutions, and would rather be considered wrong than take the trouble to prove himself right. Again and again I have taken a problem to him and have received an explanation which has afterwards proved to be the correct one. And yet he was absolutely incapable of working out the practical points which must be gone into before a case could be laid before a judge or jury.

Sherlock Holmes in "The Greek Interpreter"

The results present the data—the practical points—of a scientific paper; natural history is most directly recorded in the results section. Be thorough in your descriptions. To lay your case before the scientific jury, you should present sufficient details in the results for others to draw their own inferences and to construct their own complete explanations, and this means including a complete set of data with a fairly comprehensive indication of the variation that you found.

TABLES

For the full summary of your observations, think in terms of a table. Ideally, you can build an actual table (or a chart or a graph) in which the results are set into a simple interconnected classification scheme, but when this is difficult and when you must summarize the data in narrative form, write from a simple logical outline that resembles a table.

SELECTED CASES

At first glance, it might seem that the results would be the most objective part of a scientific paper, but all science is subjective and the results are no exception. It is not only that "our morning eyes describe a different world than do our afternoon eyes, and surely our wearied evening eyes can report only a weary evening world" (John Steinbeck, *Travels with Charley*). For a scientific paper, we must consciously select what to report—for each paper we choose a particular focus and we then try to present those data that most directly illuminate the main point. Even here we cannot, of course, tell the world everything that we have observed, so we filter out the extraneous experiences, the half-completed studies, and the equivocal observations. While you should attempt to summarize all of the relevant observations that you have amassed, you can usually present only selected cases in any detail.

How do you decide what to include and what to leave out? A good rule is: Include sufficient primary details for the reader to confidently build his own hypotheses, that is, give enough raw data for the armchair scientist to formulate his own explanation. Here, you need to present more than the predigested information of a table, because the reader needs complete data with all of their imperfections. Include at least a few full case examples so that the armchair scientist can directly see how rough are the edges of individual observations and also how variable is the data set.

One of the implicit undercurrents that wash through scientific papers is credibility. The reader's reaction to a paper is always colored by the confidence that he has in the details of the data, and thorough descriptions in which the imperfections of the real world are apparent give an important solidity to a paper.

BEST-CASE EXAMPLES AND
REPRESENTATIVE EXAMPLES

The stochasticism of our complex world insinuates variation in all but the simplest and the purest of processes, and your particular observations will undoubtedly span a range. In selecting your examples, there are two equally acceptable routes: you can present either best cases or representative cases.

"Best cases" are those observations that most clearly exemplify some ideal that you have chosen as a standard, and best cases fall at one extreme of the spectrum of actual results. For example, suppose you wish to show that axons can grow in paths that are so convoluted that they appear to be random. The actual observations may fill a spectrum from complex and convoluted to almost straight, but for your paper it is perfectly acceptable to describe only one end of the spectrum in detail, as long as you clearly state how you have selected the particular cases that you are describing. Simple phrases can indicate that you are giving best-case examples: "Axons grew in a variety of paths. Approximately 10% of the axons grew in paths that were so convoluted as to appear random, and these paths were characterized by," or "Some of the axons grew in extremely complex paths that appeared to be almost random. A good example of these was the axon designated as A6, which . . . "

On the other hand, you may choose to describe "representative cases," those cases that fall in the center of the spectrum of actual observations, and this usually means modal cases—the most common observations. Again, a well-chosen word or two indicates that you are presenting representative cases: "Although it was found that axons could grow in paths of almost any shape, most axons grew in relatively straight lines. For example, axon A6 in this series," or "Of the 46 axon paths recorded in detail, approximately half of them appeared to be relatively straight, with only bends of large radius of curvature. Axon A6 is a representative example."

EXAMPLE

The results are a logical and thorough report of the observations. The style should emphasize simple transparent descriptions, and the overall organization should follow a straightforward table-like out-line, with each paragraph expanding a separate section. For my paper, the outline is:

   I. Description of axon growth
     A. Overall appearance
     B. Growth cone
     C. Distal axon tip
  II. Quantifying the straightness of growth
     A. Fractal dimensions in general
     B. Fractal dimensions of limiting cases: random walks and straight lines
     C. Fractal dimensions of axon growth in tissue culture

RESULTS

<u>Description of Axon Growth</u>
     In tissue culture, the elongation of an axon is diffi-cult to perceive, although the movements of its tip are suf-ficiently fast to perceptibly change its shape.

[What tense should be used in scientific writing? One common rule is: statements for which you could cite other references and state-ments that are considered general knowledge should both be pre-sented in the present tense, while statements that are new or that

are specific to your experiments should be presented in the past
tense.]

On the other hand, with time—lapse recording, axons can
clearly be seen to elongate in spurts and to undergo fre—
quent short retractions. Although the axonal growth cone
sends off fine processes in all directions, the major elon—
gation proceeds forward and the axon itself tends to main—
tain its same orientation as it grows (Harrison, 1910; Katz
et al., 1984) (Fig. 1).

[In the results, cite references that have precedent as significant
descriptions and also references that present the same observations
in more detail.]

More than 40 axons were mapped in detail.

[Specify the number of observations.]

The exact form of the paths of individual axons varied, but
the general growth process was similar in all cases.

[In other words, the following description is of representative cases.]

The growth cones were continuously active. In our tissue
culture system, the growth cones formed long (6—10 μm)
spike—shaped filopodia.

[Quantify descriptors whenever possible.]

These filopodia shot out in all directions, and they often
formed temporary attachments to the substrate. Frequently,
the filopodia retracted suddenly, sometimes they "broke" at
sharp angles midway along their lengths, and sometimes their
tips waved about in the media.

[I could not resist the temptation to use a bit of fanciful language, but I think that in general colorful words and metaphors should be used only sparingly in the results.]

Concurrently, the edges of the growth cones also gave rise to short (2–3 μm) ruffle-shaped lamellipodia, which formed and disappeared continuously.

At its tip, a 10–20 μm length of the axon was quite active. Usually, the major growth cone was at the very distal end of the axon, but other regions of this active tip area occasionally became the major growth cone. The entire length of the active tip area continuously varied in diameter. Growth cone activity sometimes created temporary sharp bends in this terminal 10–20 μm length, but the sharp angles quickly (within minutes) dissipated and the tips usually had only shallow radii of curvature.

Quantifying the Straightness of Growth of an Axon (TABLE I)
Axons do not grow in perfectly straight lines (Fig. 1), but do they grow randomly?

[A simple question is a good device for defining classes of observations.]

To assess objectively the relative randomness or straightness of growth of an axon, I followed axons as they elongated and then applied a quantitative straightness measure, based on the fractal concept of Mandelbrot (1977 1983). (See the Appendix.) Fractal dimensions describe the complexity of curves. For curves in a plane, such as axon growth paths on tissue culture dishes, a simple estimate of the fractal dimension can be constructed to range between 1 (for a perfectly straight path) and approximately 2 (for an ideal random path). (This measure is outlined in the Materials and

Methods above and is described in detail in: Katz and
George, 1985.)

[Cross-referencing helps the reader, especially when he is attempting
to reconstruct a new or complicated form of data analysis. Because
scientific papers are not literary essays, you should not be afraid to
break the narrative flow with facts, definitions, and cross-references
that may be useful to the reader. Scientific papers are reference
works, and they should tend toward the encyclopedic. In this regard,
remember that sentences should always be crystal clear, even at the
expense of their smoothness.]

First, I verified that my formula for fractal dimensions (see Materials and Methods above) did indeed exhibit
these theoretical limits.

[Use the active voice whenever possible, and do not shy away from
first person singular pronouns when they are appropriate.]

Using a computer, I simulated a variety of axon paths. Axon
paths that were straight lines always had a fractal dimension $D = 1$, and, for axon paths generated as random walks in
a plane, the average fractal dimension was $D = 1.83$ (std.
dev. $= 0.301$, range $= 1.13$ to 3.20, mode $= 1.75$, $N = 501$
simulated axons each followed for 19 ten min time
intervals).

[When the results depend on specific numerical values, list the number of data points and include an exact indicator of the range of
variation.]

Next, I applied the formula to actual tissue culture
data. (To yield statistically useful fractal values, the
number of time intervals for each axon must be 8 or more
(Katz and George, 1985).) For frog axons, the average fractal dimension was $D = 1.28$ (std. dev. $= 0.223$,

range = 1.01 to 1.84, mode = 1.22, N = 23 axons followed
for an average of 14 ten min time intervals). For chick axons, the average fractal dimension was D = 1.31 (std.
dev. = 0.229, range = 1.09 to 2.00, mode = 1.25, N = 17
axons followed for an average of 21 ten min time intervals).
These fractal dimensions suggest that real axon growth paths
are quite straight and are clearly not random walks in a
plane.

    Sample fractal dimensions are lognormally distributed
(not normally distributed); thus, standard statistics must
be done on the logs of the fractal dimensions (Katz and
George, 1985). t-tests on the logs of the data demonstrated
that the straightness of growth of the frog and of the chick
axons were statistically identical (P < 0.65) and that neither type of axon grew randomly (P < 0.001).

[A table is an ideal form for presenting the bulk of the results. Each
table should have a brief title, and tables should be numbered consecutively, as they are referred to in the text.]

TABLE I

Fractal Dimensions, D

	mean D	mode	range	N
frog axons	1.28 ± 0.223	1.22	1.01–1.84	23
chick axons	1.31 ± 0.229	1.25	1.09–2.00	17
random walks	1.83 ± 0.301	1.75	1.13–3.20	501

Differences between the mean D are not significant (P < 0.65, t-
test on the logs of the data) for frogs and chicks. Differences
between mean D are significant (P < 0.001, t-test on the logs of
the data) for frogs and random walks and for chicks and random
walks.

10 / Figures

Humans are visual animals, and pictures are an excellent way for us to communicate. Beyond this, figures offer special benefits to the scientist as he or she writes a paper. First, the exercise of transforming an idea into graphic form helps to make the idea precise, and a good diagram or graph can be a self-contained model in itself. Second, figures can provide an independent test of the thoroughness and the coherence of an idea because visual logic is often different from the narrative logic used in the main sections of the paper. For these reasons, I suggest that you try to construct visual representations of all of your major data sets and concepts.

In scientific papers, the most common figures are photographs, schematic diagrams, and graphs. In general, figures portraying techniques fit in the materials and methods section, figures of actual data belong in the results, and figures of synthetic ideas, abstractions, and models should be in the discussion. Photographs are usually data, and so they are most often found in the results. The introduction rarely has any figures at all. Every figure should be referred to somewhere in the text, and the figures should be numbered consecutively, in the order that they are mentioned in the paper.

Ideally, a figure and its legend will be self-explanatory, a little story in itself. Write your figure legends in complete sentences or complete phrases, and include enough information in each legend so that the figure will make sense independent of the article. It is helpful to have the first sentence in the legend summarize the figure, because this sentence behaves like a title.

Some basic rules for figure construction include: When your figure is a diagram or a photograph an indication of the scale is always important, and when possible include a standard size mark directly in the figure. For numbers, use multiples of ten: 2500 should be 2.5×10^3. For lettering within the figure, capital letters are preferred because a lower-case L can be confused with the numeral one. In graphs comparing more than one data set, use clearly different symbols, such as solid circles and open circles, and, in general, label all axes directly on the graph. (Some journals require the lettering to be done by the printer; as usual, be sure to check the instructions to contributors before doing your final artwork.) Papp's *Manual of Scientific Illustration* (see Appendix B) is a good reference for techniques of scientific artwork.

In your final manuscript, figures are often submitted as glossy photographic prints, with individual figures mounted on separate 21.6×27.9 cm (8.5×11 inch) pieces of stiff cardboard. (Some journals require unmounted photographs or even the original; again, always check the instructions to contributors.) As a rule, black-and-white line drawings and gray tone (halftone) photographs should not be mixed in the same figure. Attempt to make the prints the exact size that they will appear in the journal; the size of the lettering and numbering in the figure can then be carefully planned in advance. (In addition, this will avoid reductions or enlargements by the printer, which can reduce the quality of the final figure.) Put a label on the back of each figure, indicating the figure number, the authors' last names, and which side of the figure is the top.

GRAPHS

Graphs translate arithmetic into geometry, transforming numbers into simple spatial patterns. As with all other parts of a scientific paper, graphs have a standard form, and it is important to stay within the conventional bounds, because an exotic or an idiosyncratic graph is difficult to read and evaluate.

Each graph should be simple, clear, and built around only one type of data. A graph should have only one set of axes (an *x*-axis and a *y*-axis), and each axis should have only one scale. Within these constraints, the background layout of the final figure can include additional standard lines and markings to help the reader. The *x*-axis can be mirrored on the top of the graph and the *y*-axis on the right—making the graph into a box—and internal standard lines, such as a 45° line, can be lightly drawn through the middle.

Try to scale the data to fill the graph, but choose units and intervals that are natural for the data, that is, units for which your audience has an intuitive feel. When graphing the numbers of spots on the back of a newt, mark your axes in natural units, whole integers—fives, tens, twenties—not in unnatural units such as fractions. Indicate five to ten standard intervals along each axis and use evenly spaced intervals, because it is difficult to interpolate approximate values between unevenly spaced intervals (such as logarithmic units) by eye.

A graph serves two purposes. First, it summarizes a set of data in spatial form. Second, taking advantage of our well-developed intuitive ability to ferret out all manner of spatial patterns, a graph lets us directly recognize patterns inherent in the data. As a summary of data, a graph need not substitute for a detailed table: you do not have to be able to read the exact numerical values of the data from the graph. Instead, the relations between the data should be accurate and the broad sweep of the data should be clear. You can help the reader to see these relations by connecting sequential points and by using the same symbol to plot related data points.

As a device for allowing us to see patterns among the data, a graph is a two-edged sword. On the one hand, spatial patterns in the data will jump out at us. On the other hand, we tend to see such patterns as if they are homogeneously laid out before us, and, when the graphic format bends, folds, twists, or clusters the data, we will often distort or misrepresent the real patterns. Logarithmic plots, for instance, downplay differences between large values; we may not be able to perceive trends hidden in the high end of logarithmic graphs, or we may overemphasize variations exposed at the low end of logarithmic graphs.

Even homogeneous graphing paradigms are not all equally useful. We cannot easily discern spatial patterns that are compressed by the scale of the numbers, and we often overlook detailed patterns that are hidden within other broader patterns. Changes in the slope of a linear graph, variations in the spacing between two lines, and relative areas under curves can be difficult to judge. Periodic variations in the intervals between data can be missed, especially when the data undergo dramatic contortions. Nonexistent periodicities form shadows in dense graphs of complex data, whereas other patterns can be lost in the tangles of such graphs. You cannot entirely protect your reader from such problems, and, although you should try your best to present clear and simple graphs composed of direct data, other nonvisual numerical analyses (such as statistical tests) should accompany your graphs to document any conclusions about patterns in your data.

EXAMPLE

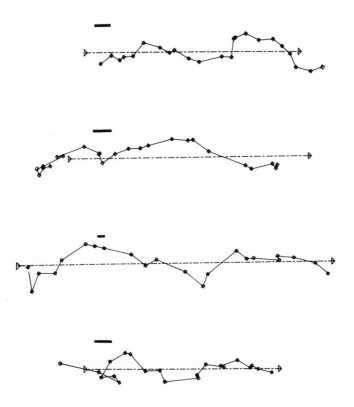

Figure 1. A computer reconstruction of the growth of four typical axons in tissue culture, indicating their tendency to grow in fairly straight paths.

[The first phrase is like a title summarizing the entire figure.]

Each small square is the location of the axon tip (the
growth cone) at the end of a successive 10 min time interval. The dashed line with arrows indicates the principal
component of growth (Sokal and Rohlf, 1969). The top two axons are embryonic frog axons, and the bottom two are embryonic chick dorsal root ganglion axons. Fractal dimensions
for the axon growth paths are (from top to bottom):
$D = 1.06$, $D = 1.05$, $D = 1.09$, $D = 1.13$. (Straight lines
will have $D = 1$; random walks will have $D \rightarrow 2$.) Bar = 10
μm.

[Always provide an indication of the scale of diagrams and of
photographs.]

11 / Footnotes and Appendices

Scientific papers should be as complete as possible. Sometimes you will wish to include discussions or technical descriptions that are directly related to the main topic but that will be of interest to only a limited number of readers. This extra information is appropriate for footnotes or appendices. A footnote is a short, one paragraph appendix.

Keep the footnotes and the appendices to a minimum. To justify the inclusion of a footnote or an appendix in your paper, it must contain information that is required for certain uses of the paper. For example, to apply a formula, readers need only the equation; but to modify or to extend the formula, readers may also need an explanation of its derivation, and here a footnote or an appendix might be in order. Use footnotes and appendices (as opposed to the main text) when the information is sufficiently specialized that the remainder of the article is self-sufficient, and when most readers can use the main article without the footnotes and appendices.

EXAMPLE

My sample manuscript uses a new and specialized mathematical analysis of the growth paths of axons. In the main text, I briefly

describe the essential formula. To further explain the meaning and
the derivation of the mathematics, I include a footnote or an appendix, and because the explanation requires more than one paragraph I will use an appendix.

APPENDIX

Topologically, a curve in a plane always has a dimension of one, but, in another sense, as it becomes more and
more convoluted, the curve fills more and more of the plane.
From this perspective, a convoluted curve can be considered
to have a dimension--a fractal dimension--of greater than
one (Mandelbrot, 1977 1983). In terms of fractals, a
straight line has a dimension of one, an irregular line has
a dimension of between one and two, and a line that is so
convoluted as to completely fill a plane has a dimension ap-
proaching the dimension of the plane, namely a dimension of
two. Fractal dimensions assign numbers to the degree of con-
volution of planar curves.

The general form of a fractal dimension D of a planar
curve is:

$$(\text{length})^{1/D} = K \ (\text{area})^{1/2}$$

where "length" signifies the total length of the curve,
"area" is the maximum potential area that the curve could
fill, and K is a constant (Mandelbrot, 1977 1983). A more
specific and practical form of this equation for planar
curves constructed of n connected line segments is:

$$D = \log(n)/\log(nd/L)$$

$$= \log(n)/(\log(n) + \log(d/L))$$

where: d is the planar diameter of the curve (here estimated
as the greatest distance between any two endpoints along the
curve) and L is the total length of the curve (the sum of the

lengths of the line segments). This formula has the following limiting values:

a. When the curve is a straight line, L is equal to the
planar diameter of the curve and the fractal dimension
D = 1.

b. When the curve is a random walk, L will usually be
approximately equal to the square root of n (the number of
steps) times the planar diameter of the curve, and the fractal dimension D --> 2.

Empirically, the lognormal distribution appears to be a
good representation for a sample population of fractal dimensions, and this means that the logarithms of the fractal
dimensions will be approximately normally-distributed.
Thus, the usual statistical analyses (e.g., t-tests and chi-
square tests) can be directly applied to the logs of the
fractal dimensions. For standard statistical analyses, use
the mean of the logs of the fractal dimensions and the variance (or the standard deviation) of the logs of the fractal
dimensions. (Further details can be found in: Katz and
George, 1985.)

12 / Discussion

"The time has come," the Walrus said,
   "To talk of many things;
Of shoes—and ships—and sealing wax—
   Of cabbages—and kings—
And why the sea is boiling hot—
   And whether pigs have wings."

Lewis Carroll,
*Through the Looking-Glass
and What Alice Found There*

A discussion analyzes the results. The discussion should be a brief
essay essentially sufficient unto itself; therefore, start with a few
sentences that summarize the most important results. Focus the
remainder of the discussion around only two or three points and
make each of these into individual analyses in separate sections.
Begin each section with a statement of your observations, add a
brief summary of relevant observations from other studies, and end
with a specific hypothesis, implication, or conclusion. In other words,
each section should proceed from the most specific (your results),
through the more general (other researchers' results), to the "uni-

versal" (general inferences and predictions). Likewise, the overall discussion should also follow this natural flow of presentation: the earlier sections should analyze particulars, such as underlying mechanisms and detailed explanations, while the last section should consider the most general and wide-reaching implications of your observations.

There is no one right way to analyze your results. Nonetheless, it is helpful to begin by organizing your discussion around three questions:

1. *The summary.* What are the major patterns among the observations?
2. *The analytic inferences.* What are the likely causes (mechanisms) underlying these patterns?
3. *The synthetic inferences.* What are the resulting predictions?
    a. Seeing these particular patterns in your system, what patterns would you predict in other systems?
    b. Inferring these particular causes for your system, what causes would you predict for other systems?
    c. Knowing of similar patterns or similar causes in other systems, what links can you suggest between these systems and yours?

Write a complete and closely reasoned analysis; and, in the end review your discussion critically, paring it down to the most simple and direct logic. It is the discussion in a paper that most often tends to wander, and it is the discussion that reviewers usually ask to be trimmed. All contemporary journals aspire to a lean and concise style, but some dictate brevity, and the discussion bears the brunt of these Spartan strictures. Although the *Journal of Comparative Neurology* states: "Generally, papers that will occupy not more than 18 to 26 printed pages are preferred, but in special cases longer articles will be accepted," the *Proceedings of the National Academy of Sciences (USA)* warns that "articles should be as brief as full documentation allows. They may not exceed five printed pages (approximately 5000 words)." And, *Science* curtly admonishes: "Length limit. Up to 2000 words (approximately 1 to 1½ printed pages), including references and notes and figures and table legends."

EXAMPLE

The discussion should make a small number of specific points. Here,
I make three points.

DISCUSSION

    Detailed measurements confirmed the impression that ax-
ons do not grow randomly, even in homogeneous environments,
and the average fractal dimension of D = 1.28–1.33 showed
that axons tend to grow straight (Fig. 1 and TABLE I).

[Begin with a restatement of the most important result.]

The determination of a specific objective number--the fractal dimension--that characterizes the relative straightness
of growth of an axon now permits a quantitative assessment
of the effect of various experimental perturbations (e.g.,
variations in the substrate adhesivity and the use of cell-
motility disrupting drugs such as taxol and cytochalasin) on
this intrinsic tendency for straight growth (Katz and
George, 1985).

[The first point of this discussion is the general use of the techniques.]

    Although growth cones tend to actively alternate sides,
axons do not grow in convoluted or "wiggly" paths. Growth
cones change their directions of movement significantly more
often than would be expected by pure chance, and the angles
through which they move span the entire range of a full 180°
(Katz, 1985). This broad searching behavior of the growth
cone appears to be strongly tempered by the axon proper, because the axon itself tends to maintain its same orientation

(to continue pointing in the same direction) throughout its
growth (Katz, 1985).

[The present tense is used here, because the information is cited
from the extant scientific literature and is, therefore, public
knowledge.]

One possible cause for the orienting and the damping
effects that the axon proper has on its route of growth is an
intrinsic resistance to bending: the average axonal bend is
only 17°, and, 95% of the time, axons are bent less than 65°
from straight lines (Katz, 1985). Axons behave somewhat like
elastic rods (Katz, unpublished observations), and the lon-
gitudinally-oriented axonal cytoskeleton may be the prime
determinant of these rod-like properties of the axon (Gilbert, 1975; Morris and Lasek, 1982 1984).

[There are two special categories of nonstandard references: unpublished observations and personal communications. Unpublished
observations are data that you have accumulated. The overwhelming
bulk of your data belongs in the results; nonetheless, those special
observations in which you have complete confidence, that are necessary for building your arguments, but that are not yet complete
enough to fully report in the results may be summarized as "unpublished observations" in the discussion. Similar observations by
other researchers are referenced as "personal communications."
Unlike unpublished observations, personal communications may be
included in any section of the paper. Be certain to clear personal
communications with the sources; some journals will require written
confirmation of a personal communication from the person cited.]

In embryonic axons, the bulk of the core cytoskeleton is
formed of microtubules (Bray and Bunge, 1981; Letourneau,
1982), and microtubules have an inherent elastic stiffness
(Kamimura and Takahashi, 1981). This longitudinal rigidity
of the axonal cytoskeleton may be the major cause of the inherent tendency of axons to grow straight.

[The second point of this discussion is a cell biological interpretation of the observations.]

The intrinsic tendency for axons to grow straight confirms the notion that when axons grow in convoluted patterns extrinsic factors must be the major determinants of the final axonal configuration. As Harrison (1910, pp. 837/835) wrote:

The outgrowth of the nerve fibers . . . can be accounted for by power of growth in a straight line modified by deflection as a result of minor obstacles in the path. . . . Given a form of protoplasm with power to extend itself in a definite direction so as to form a fiber, the next step is to determine the influences which may modify the direction of its growth and produce the specific arrangement of nerve tracts found in the mature organism.

At the other end of the spectrum, certain regions of the nervous system (such as the circumferential fibers of the spinal cord (Holley, 1983)) are characterized by sheets of parallel straight axons (Katz and Lasek, 1985). In these regions, the intrinsic tendency of axons to grow straight may be one of the critical determinants of the final axon patterns.

[The third and final point of this discussion—the conclusion—is the embryological implications of the observations.]

13 / Conclusions

"I can see nothing," said I handing it back to my friend.
"On the contrary, Watson, you can see everything. You fail,
however, to reason from what you see. You are too timid in
drawing your inferences."

Sherlock Holmes in "The Adventure of the
Blue Carbuncle"

End your paper with a firm conclusion. The discussion may make
more than one point, but its last point should be the conclusion. (It
is not necessary to have a separate section titled "Conclusions"; a
good conclusion can be one or two sentences at the end of the
discussion.)

Usually, after a brief summary of the results, the discussion
proceeds from the particular to the general, from explanations of
the observations to considerations of the broader implications; thus,
the conclusion is usually a general statement. This is a good rule,
but "general" is not the only criterion for choosing a conclusion.
When you plan the organization of your discussion, step back a
moment and ask yourself what is the strongest and most memorable
statement you can make from your observations. This is the best

synthesis of your scientific efforts, and it is the idea that you wish
to leave with your reader; state it simply and clearly.

EXAMPLE

At the other end of the spectrum, certain regions of the
nervous system (such as the circumferential fibers of the
spinal cord (Holley, 1983)) are characterized by sheets of
parallel straight axons (Katz and Lasek, 1985). In these regions, the intrinsic tendency of axons to grow straight may
be one of the critical determinants of the final axon
patterns.

[When the conclusion is the last few sentences of the discussion, it
is helpful to introduce it with "Thus," "Therefore," or "In conclusion," but in my paper these adverbs are inappropriate.]

14 / Introduction

The last major section that you write should be the introduction. Constructing a scientific paper is a research project in itself, and, having discovered the conclusion, you can now go back and relate the entire effort to the broader dimensions of the scientific edifice.

The introduction is a funnel, channeling the reader from whatever region of science he normally inhabits, bringing him into your realm, and then launching him directly into your particular experiments. As you build your introduction, begin by laying out the general scientific issues, then describe what is already known and define the current frontiers. Finally, explain exactly where your experiments fit in. "To find where you are going, you must know where you are" (John Steinbeck, *Travels with Charley*).

EXAMPLE

INTRODUCTION

Within an organism, axons are found in a great many different configurations, spanning a range from straight lines to hairpin loops (Ramon y Cajal, 1911).

[Set the stage with a statement of the general phenomenon and with an entree into the classic literature.]

In vitro studies have demonstrated that the configuration assumed by an axon can be completely determined by the pattern of local extrinsic features in the substrate through which the axon grows. For example, axons will follow the contours of the most adhesive substrate, even when those contours have acute bends (Weiss, 1945; Letourneau, 1975a; Collins and Garrett, Jr., 1980). Likewise, in vivo studies are also consistent with the idea that the configuration of an axon within an organism can be almost entirely determined by extrinsic factors, especially the contours of its local environment (e.g., Ramon y Cajal, 1911; Speidel, 1933; Rakic, 1971; Katz and Lasek, 1979; Silver and Sidman, 1980; Katz, 1984).

[Provide sufficient sources for the reader to educate himself in the details of recent observations and of contemporary thoughts across the full range of the field.]

At the same time, a number of in vitro and in vivo stud-

ies have suggested that in some situations factors intrinsic
to the neuron may also mold the configuration of its axon
(Bray, 1973; Strassman et al., 1973; Heacock and Agranoff,
1977; Lasek et al., 1983; Katz, 1984; Katz and Lasek, 1985).

[Next, begin to lay out the general problem.]

If a neuron could be grown in an ideal homogeneous
environment—an environment with no extrinsic guiding contours—then its axon should assume a configuration that was
largely determined by intrinsic factors.

[Now, describe which general experimental paradigms offer solutions.]

Tissue culture systems can provide relatively homogeneous
environments, and by growing axons in dispersed culture on
homogeneous surfaces one may be able to reveal the intrinsic
determinants of axon configurations.

From such tissue culture studies, Bray (1979) and others have shown that an important intrinsic factor determining the final axon configuration is a straightening
tendency. Due to internal tensions, axons tend to form the
shortest straight lines between attachment points. For this
reason, established axon cultures are usually composed of
straight segments of axon. Beyond this elastic straightening, however, it is possible that axons actually grow in
relatively straight paths (Harrison, 1910; Bray, 1973). The
present study follows axons as they are actively elongating
and attempts to rigorously document the extent to which axons have an inherent tendency to grow straight.

[Proceed from the general to the particular—the opposite direction
of argument presented in the discussion—and end the introduction
with a simple statement of the one most important specific question
that you will address in your paper.]

15 / Abstract and Title

Each scientific paper has an abstract or a summary that recounts the most salient facts and conclusions and that can be read independent of the rest of the paper. The abstract should be a guide to the reader by pointing out major themes and foreshadowing the logic; it is a portable microcosm of your work.

When constructing your abstract, aim for one paragraph of less than 300 words. Use no abbreviations, use no jargon or other specialized words, and rarely cite references. The abstract should not list all of the results; instead, restate only the one or two most important observations. An abstract has absolutely no literary pretensions: it is simply an outline of your essential argument in pithy narrative form.

A scientific paper is not artistic prose, and nowhere is this distinction made more clearly than in the fact of the abstract. A scientific paper can be summarized, recapitulated, and reviewed; a work of art cannot. You can capture Einstein's special theory of relativity in a textbook, and you can understand it without ever having read the original article, but Bach's *Well-Tempered Clavier* must be heard to be understood. The essence of the music permeates its form, and to know Bach's ideas you cannot rely on a synopsis: you must always return to the original "article" when it is a work of art.

EXAMPLE: ABSTRACT

ABSTRACT

Detailed growth paths of embryonic frog and chick axons were measured as the axons elongated in dispersed cultures on acid-rinsed glass surfaces.

[State what was done. Use the past tense.]

Mathematical analyses demonstrated that under these conditions axons did not grow randomly but tended to grow straight.

[State the major result. Again, use the past tense.]

It appears that an axonal resistance to bending may be the cause of the intrinsic tendency for relatively straight axonal growth.

[Present one major explanation. The present tense is a clue to readers that you are offering a general explanation.]

The natural straightness of axonal growth may be an important developmental determinant of certain in vivo axon patterns.

[Point out one significant implication. The present tense indicates a generalization.]

EXAMPLE: TITLE

HOW STRAIGHT DO AXONS GROW?

[Together, the title and the abstract form a surrogate for the paper. Rather than summarize the abstract, use the title to further shape the organization of your argument and to orient your readers. Titles are usually used for indexing articles, and it is a good plan to include key words. One technique for building a title is to list three or four terms that embody the most vital concepts of the paper and then to arrange these words into a complete phrase. Titles should be brief, less than two lines. Do not use long, multiple modifiers like "randomly growing tissue culture axons."

In addition to a title, journals frequently ask authors for a list of key words as an aid for indexing. Recall the terms you used to hunt up relevant literature in your library searches and then choose three to five appropriate words or phrases that are not used in your title. Hierarchical phrases are especially useful. For example, "regeneration, thyroxine effects on" is better than either "regeneration" or "thyroxine." For my article, I might offer these key words (arranged alphabetically and separated by slashes):]

cell behavior, quantification of / cell motility mechanisms / development of neural patterns / fractal dimensions / tissue culture, neuronal

Michael J. Katz
Neurobiology Center and Department of Biometry
Case Western Reserve University
Cleveland, OH 44106

[Follow the title with a list of the authors and with a complete address of the institution with which the work was associated. Frequently, science is a cooperative enterprise; and when more than one person has made a significant contribution to a research project, the resulting paper can have more than one author. The first author is the researcher who takes responsibility for the entire work and who has written most of the paper.]

16 / References

Scientific papers must be amply documented to tie them into the rest of the constructure of science, and the references document your work in two ways. First, the references section lists the sources for details of those observations that are only summarized in your article. Second, the references outline the historical context (an important determinant of the meaning) of both the observations and the conclusions that you present.

As your research progressed, you accumulated a set of essential references. These initial references included: descriptions of and sources for various experimental techniques, documentation of similar results, and discussions of related terminology, concepts, and theories. When you write your paper, expand and complete this basic set of references.

A complete set of references is crucial; the reader must be able to see the scope and the depth of the available relevant scientific literature. In your paper, try to document fully any statement that you make and include citations that describe the limits of your statement (the qualifications and the exceptions) as well as citations that support your contention. As background, refer mainly to those works that you and others consider to be seminal.

On the other hand, you have the responsibility of being selective.

Include only references that are related to the subject at hand; and even then do not be blindly inclusive: help the reader by filtering out questionable studies, incomplete reports, tangential research, and duplicate work.

To complete your references, begin with the citations in the initial books and articles that you have at hand. Next, go to the library and draw up an additional list of candidate citations. There are three techniques for uncovering the relevant literature:

1. *Subject or key word.* Choose three or four specific words that identify your particular research and then use the major abstracts (for example, *Applied Science and Technology Index, Biological Abstracts, Chemical Abstracts, Computer and Control Abstracts, Dissertation Abstracts International, Electrical and Electronic Abstracts,* the *Engineering Index, Index Medicus, Mathematical Reviews, Physics Abstracts, Psychological Abstracts, Sociological Abstracts*) to identify related books and articles that have been published in the last five or more years.

2. *Author.* Each scientist represents a particular area of research. From your initial reference list, find those workers who have made significant contributions to your field of interest and use the Abstracts to identify books and articles that they have published.

3. *Bibliographic citation.* Seminal papers are cited by subsequent workers who are facing the same issues. Use the *Science Citation Index* to identify books and articles that cite key papers from your initial reference list.

Reference formats are the most idiosyncratic parts of scientific journals, so be sure to consult the information for contributors section of your target journal. Usually, your final reference list should be tabulated alphabetically at the end of your paper. In the text, the references are most often presented by author and date—for example, Katz, 1984a, or Lasek et al., 1983—with an additional letter after the date when it is necessary to distinguish two articles published in the same year, and with "et al." substituting for names when there are more than two authors. Occasionally, journals number the references in the text and then index the numbers to the reference list. Copy the format in the journal that you have chosen. Standard sources for the abbreviations used in references include:

    1. *Chemical Abstracts Service Source Index*. American Chemical Society, Washington, DC.

    2. *List of Journals Indexed in Index Medicus*. National Library of Medicine, Bethesda, MD.

    3. *Serial Sources for the Biosis Data Base*. BioSciences Information Service, Philadelphia, PA.

EXAMPLE

As a rule, the reference list—the literature cited—should contain only material that is available in a library. In addition, you may cite articles in press (accepted and pending publication) and articles with a limited distribution (such as private printings, thesis reports, and special bulletins) if they are accessible to other scientists.

REFERENCES

[Single author articles in journals:]

Bray, D. (1973) Branching patterns of individual sympathetic
    neurons in culture. J. Cell Biol. 56: 702–712.
Bray, D. (1979) Mechanical tension produced by nerve cells
    in tissue culture. J. Cell Sci. 37: 391–410.

[Some journals require the full journal titles in their reference lists. Check the reference format in your target journal.]

[Two-author articles in journals:]

Bray, D. and Bunge, M. B. (1981) Serial analysis of microtu-
    bules in cultured rat sensory axons. J. Neurocytol. $\underline{10}$:
    589–605.
Collins, F. and Garrett, Jr., J. E. (1980) Elongating nerve
    fibers are guided by a pathway of material released
    from embryonic nonneuronal cells. Proc. Natl. Acad.
    Sci. (USA) $\underline{77}$: 6226–6228.

[Edited collation volumes:]

Faber, D. S. and Korn, H. (eds.) (1978) <u>Neurobiology of the</u>
    <u>Mauthner Cell</u>. Raven Press, NY.
[Include the city of publication for all monographs.]

[Authored books:]

Feller, W. (1957) <u>An Introduction to Probability Theory and</u>
    <u>Its Applications</u> (Second Ed.). John Wiley and Sons, NY,
    vol. I.
Hamburger, V. 1960. <u>A Manual of Experimental Embryology</u> (Re-
    vised Ed.). Univ. of Chicago Press, Chicago.
[Indicate the edition number when appropriate.]

[Theses:]

Holley, J. A. (1983) Development of the circumferential ax-
    onal pathway. PhD Thesis. Case Western Reserve Univ.,
    Cleveland, OH.
[A thesis is usually available at the university that granted the degree.]

[Separate articles with identical authors published in the same year:]

Katz, M. J. (1984a) Stereotyped and variable growth of redirected Mauthner axons. Dev. Biol. 104: 199–209.
Katz, M. J. (1984b) CNS effects of mechanically produced spina bifida. Dev. Med. Child Neurol. 26: 617–631.

[Articles accepted and pending publication:]

Katz, M. J. and George, E. B. (1985) Fractals and the analysis of growth paths. Bull. Math. Biol., in press.
[Include the journal name and any other available information for articles in press.]

[Multiple author articles in journals:]

Katz, M. J., George, E. B., and Gilbert, L. J. (1984) Axonal elongation as a stochastic walk. Cell Motil. 4: 351–370.
Katz, M. J., Lasek, R. J., Osdoby, P., Whittaker, J. R., and Caplan, A. I. (1982) Bolton–Hunter reagent as a vital stain for developing systems. Dev. Biol. 90: 419–429.

[Articles in collation volumes:]

Katz, M. J. and Lasek, R. J. (1985) How are the elemental axon patterns produced in the spinal cord? In: Perspectives of Neuroscience. From Molecule to Mind, Tsukada, Y., ed., Univ. Tokyo Press, Tokyo, pp. 43–60.
Lasek, R. J., Metuzals, J., and Kaiserman–Abramof, I. R. (1983) Cytoskeletons reconstituted in vitro indicate that neurofilaments contribute to the helical structure of axons. In: Developing and Regenerating Vertebrate

Nervous Systems, Markwald, R. R. and Kenny, A. D.,
eds., Alan R. Liss, NY, pp. 1—18.

[Always include the editors' names when referencing collation volumes.]

[Articles in supplements to journals:]

Sperry, R. W. (1951) Regulative factors in the orderly
growth of neuronal circuits. Growth, supplement to vol.
15, pp. 63—87.

17 / Acknowledgments

> This book is based . . . on its predecessors. . . . The compiler is indebted to every book he ever read and every person he ever knew.
>
> B. Evans, Acknowledgments,
> *Dictionary of Quotations*

When your paper is finished, there is one last piece to fit into the whole. After the conclusion and before the references lies a short section of acknowledgments, which is your opportunity to thank those who helped you technically, intellectually, and financially. Besides being an item of scholarly politeness, the acknowledgments serve an important scientific function. Matters of science take meaning from their historical context as much as from their intrinsic merits: the acknowledgments set your work directly into the context of your immediate scientific community, and the acknowledgments section is a historical statement of your scientific efforts.

EXAMPLE

ACKNOWLEDGMENTS

I am indebted to W. F. Eddy for many constructive con-
versations about the problems of analyzing the straightness
of axonal growth, to E. B. George for a computer program
that calculates principal components, to T. Hoshiko for reminding me about fractals, to R. J. Lasek for first raising
the question of how straight axons actually grow, and to
L. J. Gilbert and L. F. Watson for excellent technical
assistance.

[To identify individuals properly, give their full initials or their first names.]

This work was supported by grants from the National Institutes of Health and the Whitehall Foundation and by an
Alfred P. Sloan Foundation Research Fellowship.

[Always write in complete sentences.]

18 / The Art of a Scientific Paper

> "You see, my dear Watson"—he propped his test-tube in the rack and began to lecture with the air of a professor addressing his class—"it is not really difficult to construct a series of inferences, each dependent upon its predecessor and each simple in itself."
>
> Sherlock Holmes in "The Adventure of the Dancing Men"

Write the main sections of your paper like a mystery novel, and let the reader gradually discover the conclusions himself. In the introduction, set out the problem. In the materials and methods section, present your tools and techniques, set the stage and give a complete roster of the cast of characters, including the minor players. In the results, reveal your observations one by one in a complete and orderly sequence. In the discussion, build toward the conclusion from the particular to the general: first, give a one- or two-sentence restatement of the essential results as a starting post; next, add relevant observations from other studies; finally, tie the whole together, leading resolutely toward a possible explanation of the phenomenon and toward a simple statement of the implications of the observations. The conclusion is then the denouement.

A memorable scientific article makes but one clear point; at base, it should address only one issue and it should have a clear arrow. When you have finished the entire manuscript, go back to the beginning and rework the wording so that each sentence points toward the conclusion. Be bold and ruthless in trimming extraneous and tangential statements, especially in the introduction and the discussion.

Overall, a scientific paper should have the form of a mystery novel and the shape of a comfortable breath. It should slowly rise with inspiration to the beginning of the discussion and then fall unerringly toward the conclusion. Along the way, it should be paved with tiny transparent verities, "small diagnostic truths which are the foundations of the larger truth" (J. Steinbeck, *Travels with Charley*).

19 / Gestation and Rewriting

I cast thee by as one unfit for light,
Thy Visage was so irksome in my sight;
Yet being mine own, at length affection would
Thy blemishes amend, if so I could:
I wash'd thy face, but more defects I saw,
And rubbing off a spot, still made a flaw.
I stretcht thy joints to make thee even feet,
Yet still thou run'st more hobling than is meet;
In better dress to trim thee was my mind,
But nought save home-spun Cloth i' th' house I find.

Anne Bradstreet, "The Author to Her Book"

Inevitably, the first draft of a scientific article is unbalanced: ideas that capture your fancy during the writing are overemphasized, new discoveries are not thoroughly explored, and old concepts are explained too tersely. In the struggle to write a comprehensive paper, it can be hard for the author himself to sort the major discontinuities, deformities, and misplaced or mistaken constructs from the ubiquitous small blemishes, but many of the serious problems can be readily identified by colleagues—outsiders who are kind enough to read through the manuscript. Find a friendly reader. Ask him to mark sections that he must read through twice to understand, places where he finds his attention wandering, and any phrases that feel awkward. Besides the comments that he volunteers, specifically ask him to point out holes—missing data or ideas—and tangential, extraneous, unnecessary, and digressive inclusions.

When I finally finish a manuscript, I am always tired of it. The effort of reworking the paper deadens my mind, I take to absently studying the acknowledgments, and I have no doubt that I should immediately send the typescript to the journal. Moreover, I feel I have accomplished something important, and no matter how friendly and tactful my collegial reader, his critique always makes me a bit angry. I tap my feet and jiggle my pen, but usually a small rational voice reminds me that the work must be lucid and comprehensible to interested colleagues, and because of this tiny voice I do my best to listen patiently and dispassionately to the comments, suggestions, and criticisms. I thank my reader, and I set the manuscript aside until after a good night's sleep my enthusiasm returns. Then, I attack the paper afresh, armed with a new objectivity that spurs me on to write an article surely destined for a Nobel prize.

Although more difficult, it is also possible (and desirable) to be your own critic at some stage along the way. For this you must let the article sit. We all become attached to our children, and pouring time and effort into the composition of a scientific narrative builds maternal bonds, so that we love the distorted and the incomplete as well as the harmonious and the well-formed. A gestation period of at least a week, and better yet a month, takes some of the emotional edge from your writing. Wait until one quiet morning and then sit down with the intent of rebuilding the entire manuscript, starting with the title and working through the whole paper from beginning to end.

The two basic steps in revising a manuscript are filling in incomplete logic steps and cutting out tangential or weak statements. All improvements seem to fall into one of these two categories. Read each sentence and ask:

1. What is the main point, and can it be said more simply or more clearly? Is it necessary, or is it a digression that can be omitted?

2. Does it follow from the sentence before? Is the logic correct,

and is it complete? Are there hidden thoughts that must be stated explicitly?

At each point, my best counsel is: Be exacting and unemotional. If there is any question about the clarity of an argument, expand it; and, if you are at all uncomfortable about an idea, remove it.

> Behold once more with serious labor here
> Have I refurnished out this little frame,
> Repaired some parts defective here and there,
> And passages new added to the same,
> Some rooms enlarged, made some less than they were,
> Like to the curious builder who this year
> Pulls down, and alters what he did the last
> As if the thing in doing were more dear
> Than being done, & nothing likes that's past.
>
> Samuel Daniel, "To the Reader"

20 / Editors, Referees, and Revisions

You have sent off a polished version of your manuscript to a journal, and after a few months you receive a packet with an editor's letter and one or two reviews by independent referees. The editor's letter will put your paper into one of three categories: the manuscript will be accepted after some specified revisions, it will require revision and then re-review, or it will have been rejected. Regardless of the categorization, use this as an opportunity to rework your paper, and at this point I suggest that you revise the manuscript in two steps.

MAJOR OVERHAUL

The mandatory wait while your paper is being reviewed is an enforced gestation period. Before you read the referees' comments, take a moment to reread the manuscript and ask yourself: Where is this paper likely to run into difficulties, and with what am I most uncomfortable? Target these areas for thorough reworkings, irrespective of the reviewers' comments.

Now, read quickly through the reviews to discover their general drift. Do the major problems relate to specific observations (data), to logic statements (analysis), or to organization (presentation)? To which sections of the manuscript are most of the criticisms directed? When you can answer these questions, then I suggest that you set the reviews aside, take the manuscript in hand, and begin to rewrite.

SPECIFIC IMPROVEMENTS

After these large-scale changes, it is time to take advantage of the more specific comments of the referees. I find that a good device for organizing the specific improvements is the letter that I write to accompany my revised manuscript. Even if your paper is rejected, you can resubmit it with a thoughtful letter. In any case, when you return your paper to the editor, you should include a detailed list of the changes that you have made. In the first paragraph, summarize the major overhaul. Tell the editor which sections have been revised and in what ways they are now different.

The remainder of the letter should itemize the specific improvements. Take the reviews sentence by sentence, and if the suggestions make sense to you, try to make the indicated changes. With each change, write a sentence in your letter briefly stating what the referee suggested and how you have followed that suggestion. Remember that although most reviewers write with the best of intentions, they do not always have the time to produce thorough and completely well-thought-out or error-free critiques. Some of their comments will not be useful. I suggest that you just pass over these particular comments and that you try not to take even the emotional comments personally. All scientists have been hit with criticisms that sting. A referee once wrote of one of my manuscripts: "If this paper is published it will set science back 50 years." The best thing to do with such comments is to save them as anecdotes for your book on how to write a scientific paper.

APPENDIX A

This is the paper [M. J. Katz (1985) J. Neurosci. 5:589-595, reprinted with permission of The Williams & Wilkins Company] from which sections were excerpted as examples in the main text.

HOW STRAIGHT DO AXONS GROW?

(cell behavior, quantification of/cell motility
mechanisms/development
of neural patterns/fractal dimensions/tissue
culture, neuronal)

Michael J. Katz
Neurobiology Center and Department of Biometry
Case Western Reserve University
Cleveland, OH 44106

ABSTRACT

Detailed growth paths of embryonic frog and chick axons
were measured as the axons elongated in dispersed cultures
on acid-rinsed glass surfaces. Mathematical analyses demonstrated that under these conditions axons do not grow randomly but tend to grow straight. Growth cones appeared to
actively alternate sides—right and left from the straight
line of growth—and the growth cone neck exhibited all possible angles, but the axon itself maintained a fairly constant orientation. Thus, an axonal resistance to bending may
be the cause of the intrinsic tendency for relatively
straight axonal growth. The natural straightness of axonal
growth can be an important developmental determinant of certain in vivo axon patterns.

INTRODUCTION

Within an organism, axons are found in a great many
different configurations, spanning a range from straight
lines to hairpin loops (Ramon y Cajal, 1911). In vitro studies have demonstrated that the configuration assumed by an
axon can be completely determined by the pattern of local
extrinsic features in the substrate through which the axon
grows. For example, axons will follow the contours of the
most adhesive substrate, even when those contours have acute
bends (Weiss, 1945; Letourneau, 1975a; Collins and Garrett,
Jr., 1980). Likewise, in vivo studies are also consistent
with the idea that the configuration of an axon within an
organism can be almost entirely determined by extrinsic factors, especially the contours of its local environment
(e.g., Ramon y Cajal, 1911; Speidel, 1933; Rakic, 1971; Katz
and Lasek, 1979; Silver and Sidman, 1980; Katz, 1984).

At the same time, a number of in vitro and in vivo studies have suggested that in some situations factors intrinsic
to the neuron may also mold the configuration of its axon
(Bray, 1973; Strassman et al., 1973; Heacock and Agranoff,
1977; Lasek et al., 1983; Katz, 1984; Katz and Lasek, 1985ab).
If a neuron could be grown in an ideal homogeneous environment—an environment with no extrinsic guiding contours—then
its axon should assume a configuration that was largely determined by intrinsic factors. Tissue culture systems can provide
relatively homogeneous environments, and by growing axons in
dispersed culture on homogeneous surfaces one may be able to
reveal the intrinsic determinants of axon configurations.

From such tissue culture studies, Bray (1979) and others
have shown that an important intrinsic factor determining the
final axon configuration is a straightening tendency. Due to
internal tensions, axons tend to form the shortest straight
lines between attachment points. For this reason, established
axon cultures are usually composed of straight segments of
axon. Beyond this elastic straightening, however, it is possible that axons actually grow in relatively straight paths
(Harrison, 1910; Bray, 1973; Trinkaus, 1984). The present
study follows axons as they are actively elongating and attempts to rigorously document the extent to which axons have
an inherent tendency to grow straight.

MATERIALS AND METHODS

Cultures
A. Frog Neurons
Xenopus embryos were obtained from matings of adult
frogs—matings were induced by injection of human chorionic
gonadotropin (Sigma Chemical Co., St. Louis, MO). Twenty-
four to 48 hr old amphibian primary tissue cultures were

grown by disaggregating neural tube cells of tailbud stage
Xenopus embryos and plating these cells on acid-rinsed glass
coverslips set in the bottoms of 35 mm sterile FALCON petri
dishes (Spitzer and Lamborghini, 1976). [Polyornithine-
coated cover slips were made according to Letourneau (1975a)
(poly-L-ornithine, ICN Nutritional Biochemicals, Cleveland,
OH).] Stage 28-30 (Nieuwkoop and Faber, 1967) Xenopus embryos were removed from their vitelline membranes and washed
through four changes of sterile Steinberg solution, pH 7.5
(Hamburger, 1960). Next, the ectoderms were peeled off and
the brain primordia were dissected free and washed in fresh
Steinberg solution. Brain primordia were then disaggregated
by soaking them for 10 min in a calcium- and magnesium-free
medium (59 mM NaCl, 0.7 mM KCl, 0.4 mM EDTA, pH 7.5). Cell
aggregates were broken apart by pipetting them briefly at 5
min intervals with a fire-polished Pasteur pipette. Cell
suspensions were then plated in fresh Steinberg solution
supplemented with 2% fetal bovine serum, 5-10 µM nerve
growth factor (NGF), and antibiotics (50 units/ml penicillin, 0.05 mgm/ml streptomycin--Gibco Labs, Grand Island,
NY). Approximately 1-1.5 brain primordia were plated per 35
mm culture dish and grown at 21°C. Under these conditions, 24
hr cultures averaged 150-170 differentiated cells per 22x22
mm coverslip and approximately ¼ of these cells were
neurons.

B. Chick Neurons

    Twenty-four to 48 hr old avian primary tissue cultures
were grown from dorsal root ganglion cells of 7-12 day chick
embryos (Shaw and Bray, 1977). Cells were isolated by treat-
ing ganglia with trypsin (2.5 mg/ml) and collagenase (0.5
mg/ml) at 37°C for 25 min in calcium- and magnesium-free
Hank's solution and then disaggregating the cells by passing
them through a small-bore pipette. Cultures of isolated
cells were enriched for neurons by differential adhesion:

First, the cells were grown for 30—40 min in Leibovitz's L—
15 medium supplemented with 0.6% glucose and antibiotics.
Next, the medium was replaced with complete culture medium
and the dish was agitated on a vortexer to free the loosely
adherent neurons. The resultant cell suspension was plated
on acid—washed glass coverslips that had been sealed over a
1.5 cm hole drilled in the bottom of a 35 mm culture dish.
Finally, fresh culture medium was added. The complete culture medium consisted of Leibovitz's L—15 medium (Gibco
Labs, Grand Island, NY), with 10% fetal bovine serum, 0.6%
glucose, 0.3% methyl cellulose, 5—10 μM NGF, 100 units/ml
penicillin, 0.1 μg/ml streptomycin, and 0.3 μg/ml cytosine
arabinoside added.

Data Collection

A. Criteria for Choosing Axons

Axons selected for study were only from clearly healthy
cultures and were in areas of minimal debris. The axons were
all at least 100 μm long and were separated from other cells
and cell processes by at least 100 μm. Data were analyzed
only for axons that grew without major retractions and without contacting other cells.

B. Definition of Standard Axonal Sites

At the end of each time period, coordinates were measured at five sites along each axon (Figs. 1 & 2). The growth
cone site was defined to be the center of the palm—like
growth cone. The neck site was taken to be the constriction
or the point of sharp angle at the proximal end of the growth
cone. (A clear neck point could only be identified about
half of the time.) Three additional sites were located more
proximally along the axon, using the growth cone site as the
initial point of reference: point A was taken to be 20 μm
along the axon proximal to the growth cone, point B was
taken to be 30 μm along the axon proximal to the growth cone,

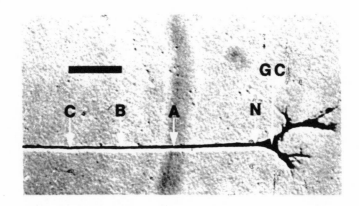

Figure 1. Micrograph of fixed chick DRG axon in culture, indicating sites of measurement—cf. Fig. 2. (Bar = 10 μm.
Bodian stain with intensifier, see: Katz and Watson, 1984.).

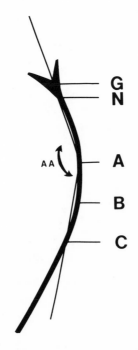

Figure 2. Drawing of terminal 50 μm of an axon indicating
five standard sites (G, N, A, B, C) of measurement during
quantification of axonal growth. AA = angle of bend of terminal 40 μm of axon. See text for details.

and point C̲ was taken to be 40 μm along the axon proximal to
the growth cone.

C. Measurement Procedures

Axonal growth was monitored at 400x magnification on an
Olympus IMT inverted phase microscope to which was attached
an RCA TC2000 Newvicon video camera equipped with an Olympus
FK 3.3x projection eyepiece. Time–lapse video records were
made on SONY BR video cassettes (¾" tape) using an NEC video
recorder (VC–9507) running at ¹⁄₆₄th real time. Real time was
indicated with a VICOM V240 date/time generator. The locations of the standard axonal sites were measured on a transparent grid that overlay the video screen (PANASONIC WV–
5300). The standard units of measurement were marked on a
flexible plastic template which was placed on the screen in
order to accurately measure distances along the curved axons. The screen had a diagonal length of approx. 21 cm, the
grid was divided into 5 squares/cm, and in our system this
resulted in a scale of 2.56 μm/square for the final video
image. Data were collected by stopping (freezing) the tape
at an exact time point and then measuring the simultaneous
locations of the five standard axonal sites. Figure 3 presents reconstructions of the growth cone movements of four
axons recorded in this manner.

D. Precision of Measurements

A static image (a calibrated stage micrometer) was recorded for two hours, rotated 90°, and then recorded for an
additional two hours. The image did not drift, and distances
remained undistorted by the rotation. Locations could be
read from the screen to an accuracy of ⅓ grid square (approx. 1 μm). However, by repeating readings for 80 separate
time points on data tapes of growing axons, we found our
overall average error to be closer to 2 μm.

Computations

A. Straightness of Growth

As a standardized measure of the straightness of the
path of growth of an axon, I used a variant of Mandelbrot's
(1977 1983) elegant fractal measures. (Fractals have been
used to characterize the windiness of rivers and of coast-
lines.) The fractal dimension D is the fractional dimension
of a curve--here, it is the fractional dimension of the path
of the axon in the plane. The formula used is:

$$D = \log(L/a)/\log(d/a)$$

where: D is the fractal dimension, L is the total length of
the path of the axon (the sum of the distances traveled in
all of the 10 min intervals), d is an estimate of the largest
distance actually covered, and a is the average length of a
step or an observed interval of growth. (d should be a diameter of the potential area covered by a random path. Here,
it is estimated by taking the largest distance between any
two points on the axon's growth path.) In this way, a completely straight growth path gives a fractal dimension
D = 1, and a completely random walk gives a fractal dimension D→2. Statistics should be done on the logarithms of the
fractal dimensions. [Further details, including a computer
program for all calculations, can be found in Katz and
George (1985).]

B. Change in Direction of Growth Cone Movement

The overall direction of axon growth during a 10 min
time interval was calculated by simple trigonometry from the
(x,y) coordinates at the beginning and at the end of the
time period. The direction was expressed as an angle in relation to an arbitrary but fixed axis, and the change in direction of growth was then the change in this angle between
successive 10 min intervals. To reveal any tendency for the
axons to turn to one particular side, right-sided angle

changes were arbitrarily called ''positive'' and left—sided
angle changes were called ''negative.''

C. Change in Axonal Orientation

The orientation of the end of an axon was determined at
the beginning and at the end of each 10 min time interval in
relation to an arbitrary but fixed axis. The locations of
the growth cone site and the neck site (or point $\underline{A}$ when the
growth cone neck had no bend) were recorded as (x,y) coordinates, and then the orientation of the end of the axon was
computed as the arctangent of the difference in the y values
divided by the difference in the x values. (A computer program containing this calculation, as standardized to an arbitrary axis, is published in: Katz et al., 1984.) The
change in orientation was the change in this angle during a
10 min period. To reveal any tendency for the axons to turn
to one particular side, right—sided deviations were arbitrarily called ''positive'' and left—sided deviations were
called ''negative.''

D. Axonal Bends

The overall bend in the terminal 40 μm of an axon was
computed for many axons at 10 min intervals. Three distances
were used in this calculation: the distance $\underline{GA}$ between the
growth cone and point $\underline{A}$, the distance $\underline{AC}$ between point $\underline{A}$ and
point $\underline{C}$, and the distance $\underline{GC}$ between the growth cone and
point $\underline{C}$. The overall bend at point $\underline{A}$ was taken to be the angle $\underline{AA}$ (Fig. 2), where:

$$AA = 2 \arctan(R/(S-GC))$$

and where:

$$S = (1/2)(GC+AC+GA)$$
$$R = ((S-GC)(S-AC)(S-GA)/S)^{1/2}$$

(This is the standard trigonometric formula for calculating an
angle of a triangle when given the lengths of the three sides.)
To reveal any tendency for the axons to bend to one particular
side, right-sided bends were arbitrarily called "positive" and
left-sided bends were called "negative."

E. Neck Bends

  The overall bend at the neck of the growth cone was computed
in a similar manner for many axons at 10 min intervals. Here, the
three distances used in the calculation were: the distance between the center of the growth cone and the neck of the growth
cone, the distance between the neck and point $\underline{A}$, and the distance
between the growth cone and point $\underline{A}$. When there was no acute bend
at the growth cone neck, it was not always possible to unequivocally distinguish the neck point. In these cases, the radius of
curvature of the neck region was approximately the same as the
radius of curvature of the terminal 40 μm of axon and the overall
bend in the latter region was taken as an approximation of the
bend at the neck region.

RESULTS

Description of Axonal Growth

  In real time, the elongation of an axon is difficult to
perceive, although the growth cone movements are sufficiently fast to perceptibly change the shape of the axon
tip. In time lapse, on the other hand, the axon can be seen
to elongate in spurts and to undergo frequent short retractions. Although the growth cone sends off microprocesses in
all directions, the major axonal elongation proceeds forward
and the axon tends to maintain its same orientation (Fig. 3)
(Katz et al., 1984).

  The growth cone is continuously active. In our tissue

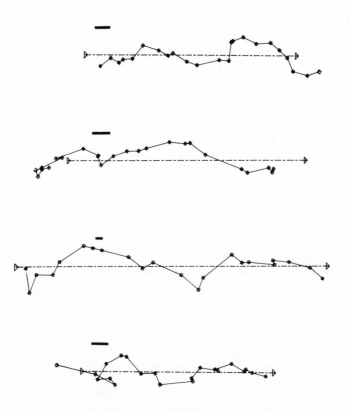

Figure 3. Computer reconstruction of growth paths of four typical axons in tissue culture, indicating their tendency to grow in fairly straight paths. Each square is the location of the growth cone at the end of a successive 10 min time interval. The dashed line with arrows indicates the principal component of growth (Sokal and Rohlf, 1968). The top two axons are embryonic frog axons; the bottom two axons are embryonic chick DRG axons. Fractal dimensions for the axons are (from top to bottom): D = 1.06, D = 1.05, D = 1.09, D = 1.13. Bar = 10 μm.

culture system, the growth cones form long (6–10 μm) spike-shaped filopodia, which can shoot out in any direction and which often form temporary attachments to the substrate. Frequently, these filopodia retract suddenly, and sometimes

they ''break'' midway along their lengths at sharp angles,
waving their tips in the media. Concurrently, the edges of
the growth cone also give rise to short (2–3 μm) ruffle-
shaped lamellipodia which form and disappear continually.

A 10–20 μm length of the axon tip is quite active and
continually varies in diameter. The major growth cone is
usually at the very distal end of the axon, but other regions of this active tip area occasionally give rise to the
major growth cone. Growth cone activity can create temporary
sharp bends in this 10–20 μm length, but the sharp angles
quickly (within minutes) dissipate and most of the time the
tip has only a shallow radius of curvature.

Quantifying the Straightness of Growth of an Axon (TABLE I)

Axons do not grow in perfectly straight lines (Fig. 3),
but do they grow randomly? To assess objectively the relative randomness or straightness of growth of an axon, I followed axons as they elongated and applied a quantitative
straightness measure, based on the work of Mandelbrot (1977
1983). One of Mandelbrot's measures, the fractal dimension,
describes the complexity of a curve. For curves in a plane,
such as axon growth paths in tissue culture, the fractal dimension D can be made to vary between 1 (for a perfectly
straight path) and approximately 2 (for a completely random
path).

First, I verified that my formula for fractal dimensions (see ''MATERIALS AND METHODS'' above) did indeed exhibit these theoretical limits. A variety of axon paths were
simulated on a computer. Straight line axon paths always had
a fractal dimension $D = 1$, and, for axon paths generated as
limited random walks in a plane, the average fractal dimension was $D = 1.83$ (std. dev. $= 0.30$, range $= 1.13$ to 3.20,
$N = 501$ simulated axons each followed for 19 ten min time
intervals).

TABLE I

Fractal Dimensions, D

	mean D	mode	range	N
frog axons:				
glass	1.28 ± 0.22	1.22	1.01–1.84	23
polyornithine	1.33 ± 0.18	1.31	1.14–1.72	11
chick axons	1.31 ± 0.23	1.25	1.09–2.00	17
random walks	1.83 ± 0.30	1.75	1.13–3.20	501

Differences between the mean D are not significant
(P < 0.65, t–test on the logs of the data) for frogs and
chicks. Differences between the mean D are significant
(P < 0.001, t–test on the logs of the data) for frogs and
random walks and for chicks and random walks.

    Next, I applied the formula to actual tissue culture
data. [To produce statistically useful sample fractal dimensions, the number of time intervals for each axon must be 8
or more (Katz and George, 1985).] For frog axons, the average fractal dimension was D = 1.28 (std. dev. = 0.223,
range = 1.01 to 1.84, N = 23 axons followed for an average
of 14 ten min time intervals). For chick axons, the average
fractal dimension was D = 1.31 (std. dev. = 0.229,
range = 1.09 to 2.00, N = 17 axons followed for an average
of 21 ten min time intervals). The fractal numbers show that
real axon paths are quite straight and are not random walks
in a plane. Sample fractal dimensions are lognormally distributed (not normally distributed); thus, standard statistics must be done on the logs of the values (Katz and George,

1985). t-tests on the logs of the data demonstrated that the straightness of growth of the frog and the chick axons were statistically identical (P < 0.65) when grown on homogeneous glass surfaces and that neither type of axon grew randomly (P < 0.001).

As a test of the effect of substrate adhesivity on the straightness of growth, I also calculated fractal dimensions for frog axons grown on polyornithine-coated acid-rinsed glass coverslips. Here, the average fractal dimension was D = 1.33 (std. dev. = 0.18, range = 1.14 to 1.72, N = 11 axons followed for an average of 18 ten min time intervals). The straightness of growth on polyornithine is statistically indistinguishable from the straightness of growth on uncoated acid-rinsed coverslips (P < 0.50, t-test).

Changes in Direction of Growth Cone Movement

For each 10 min interval of data collection, the change in the direction of growth cone movement was calculated. The mean and the standard deviations were similar for both types of axons studied. For frog axons, the mean change in growth direction was 72° (std. dev. = 56°, range = 180° to 0°, N = 337); for chicks, the mean was 77° (std. dev. = 59°, range = 180° to 0°, N = 343). The means did not differ significantly (P < 0.2, t-test), and histograms of the angle distributions for the two types of axons had similar shapes, but chi-square tests showed that the two distributions were significantly different at the P < 0.001 level.

Fig. 4 is a set of histograms of these angle distributions, when they have been assigned right and left signs (positive is right, negative is left). The values fall in symmetric curves centered at about 0°, indicating that the axons had an approximately equal probability of deviating to the right or to the left.

To assess the tendency of an axon tip to change its direction of growth in a consistent direction, I counted the

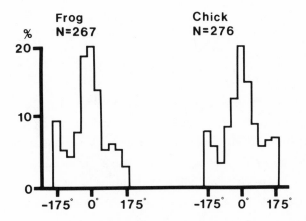

Figure 4. Histograms of changes in the direction of the growth cones of embryonic frog and chick axons grown on glass in dispersed tissue cultures. Growth cones can be found growing in any direction from the axon, but there is a tendency to grow at either moderately shallow angles or at very large angles (usually retractions—see: Katz et al. [1984]).

frequency of deviations to the same side in consecutive 10 min intervals. Then, I compared the actual numbers with expected numbers that were calculated under the assumption of independence. This assumption presumes that the change of direction in one 10 min interval is not at all correlated with the change in direction in the previous 10 min interval. The expected number is simply $N/2$—where N is the total number of consecutive time intervals. For frogs, the axons changed direction to the same side in 107 of a possible 293 times. Chi-square analyses show that this is statistically significant at the $P < 0.001$ level. Likewise, for chicks, 109 of a possible 253 consecutive changes were in the same direction, and this is statistically significant at the $P < 0.025$ level. Thus, one must reject the hypothesis that the changes of growth direction are independent in each 10 min interval. Instead, it appears that the growth cone actively alternates growth directions, because it changes its

direction of growth more often than would be expected by
pure chance.

Changes in Axon Orientation

For each 10 min interval of data collection, I calculated the change in the orientation of the end of an axon.
The means and the standard deviations differed (P < 0.001,
t-test) between the two types of axons: for frog axons, the
mean change in orientation was 23° (std. dev. = 33.5°,
range = 132° to 0°, N = 337); for chicks, the mean was 15°
(std. dev. = 16°, range = 109° to 0°, N = 343). Histograms
of the angle distributions for the two types of axons had
similar shapes, but chi-square tests also showed that the
two distributions were significantly different at the
P < 0.025 level. Another indication of this difference is
that, in our culture systems, the range of angle deviations
differed significantly between frogs and chicks. For frogs,
the change in orientation of the tip of an axon was less than
75° in >95% of the 10 min time periods. In contrast, for
chicks the change in orientation of the tip of an axon was
less than 50° in >95% of the 10 min time periods.

Figure 5 is set of histograms of these angle distributions, when they have been assigned right and left signs
(positive is right, negative is left). The values fall in
smooth bell-shaped curves centered at 0°, indicating that
the axons had an approximately equal probability of deviating to the right or to the left.

Axonal Bends

A. The Terminal 40 μm of Axon

I computed the average degree of bend in the terminal
40 μm of axon for many individual observations separated by
10 min time intervals. The results were identical for both
types of axons: for frog axons, the mean bend was 17° (std.
dev. = 20°, range = 111° to 0°, N = 299); for chick axons,

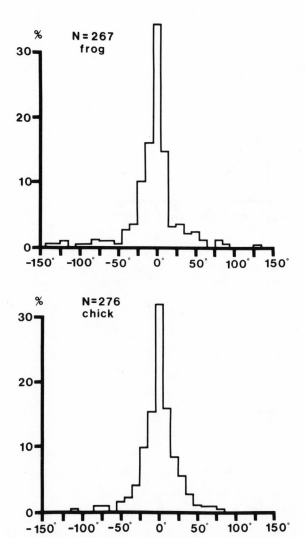

Figure 5. Histograms of changes in the orientations of embryonic frog and chick axons growing on glass in dispersed
tissue cultures. Changes in axonal orientation were much
more restricted than changes in the direction of growth cone
movement (cf. Fig. 4). Most frog axons changed their orientation less than 75° during a 10 min time interval, and most
chick axons changed their orientation less than 50° during a
10 min time interval.

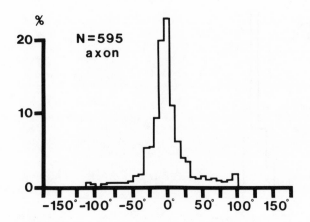

Figure 6. Histogram of the bends observed in the terminal 40 μm of embryonic axons growing on glass in dispersed tissue cultures. The average bend was only 17°, and most of the time the bend was less than 65°.

the mean bend was 17° (std. dev. = 20°, range = 116° to 0°, N = 296). In addition, histograms of the distributions of bends for the two types of axons had identical shapes, and chi-square tests showed that the two distributions were not significantly different (P < 0.10). Because the bend distributions for the frog axons could not be distinguished from the bend distributions for the chick axons, I pooled the data. For all axons, the mean bend in the terminal 40 μm of axon was 17° (std. dev. = 20°, range = 116° to 0°, N = 595). From the distribution histogram, I calculated that most of the time (>95% of cases) the bend in the terminal 40 μm of an axon was less than 65°. [A qualitatively similar observation was reported by Gundersen and Park (1984).]

The distribution of bends is plotted as a single histogram in Fig. 6, where the bends have now been assigned right and left signs (positive is right, negative is left). The values fall in a smooth bell-shaped curve centered at 0°, indicating that the axons were equally likely to bend to the right as to the left. [Gundersen and Barrett (1980) reported a similar result.]

To assess the tendency of an axon tip to continue bending in the same direction, I counted the frequency of bends
to the same side in consecutive 10 min intervals. Then, I
compared the actual numbers with expected numbers that were
calculated under the assumption of independence. This assumption presumes that the direction of bend in one 10 min
interval is not at all correlated with the direction of bend
in the previous 10 min interval. (The expected number is
simply N/2—where N is the total number of consecutive time
intervals.) For both frog axons and chick axons, axons continued bending to the same side in a disproportionately
large number of cases, and chi-square analyses showed that
this tendency was statistically significant (P < 0.01).
Thus, one must reject the hypothesis that the direction of
axonal bending is independent in each 10 min interval. Instead, it appears that, over 10 min intervals, the axon
tends to bend or to remain bent to the same side more often
than would be expected by pure chance. This result is consistent with my observations that the inherent straightening
process in an axon can take as long as 10 or more minutes to
complete (Katz, unpublished observations).

B. The Growth Cone Neck Region
    I computed the average degree of bend at the growth
cone neck for many individual observations separated by 10
min time intervals. The results differed significantly
(P < 0.01, t-test) between the two types of axons: for frog
axons, the mean bend was 37° (std. dev. = 50°, range = 179°
to 0°, N = 299); for chick axons, the mean bend was 21° (std.
dev. = 26°, range = 176° to 0°, N = 296). Histograms of the
distributions of bends for the two types of axons had similar shapes, but chi-square tests also showed that the two
distributions were significantly different at the P < 0.001
level.
    For each type of axon, the distribution of neck bends

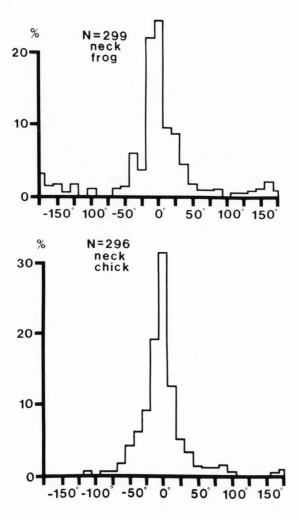

Figure 7. Histograms of the bends observed at the growth
cone neck of embryonic frog and chick axons. The range of
neck bends was significantly greater than the range of bends
in the more proximal region of the axon (cf. Fig. 6).

is plotted as a histogram in Fig. 7. For both types of axons,
the values fall in smooth bell-shaped curves centered at 0°,
indicating that the axons were equally likely to bend to the
right as to the left. From the distribution histograms, the
bend at the frog growth cone neck was found to be less than
160° >95% of the time, but for chick axons the neck bend was

found to be less than 80° >95% of the time. In other words,
in these particular culture systems, the growth cones of
frog axons tended to bend through twice the range of angles
as did chick growth cones.

Another revealing measure is the correlation between
the side to which the growth cone itself is bent and the side
to which the axon is bent. There was a clear tendency for the
growth cone and the axon to be bent to the same side, but
frequently the growth cone was bent in the opposite direction from the remainder of the axon. Specifically, the
growth cone was bent in the opposite direction in 24% of the
308 separate observations. This demonstrates that bending at
the growth cone neck can be independent from the bending of
the more proximal axon. The independence of bending in these
two regions is also seen in the statistical analyses: by
both the t-test and the F-test, the mean bend and the variance differed significantly between the two regions
($P < 0.001$ for frogs, $P < 0.02$ for chicks).

DISCUSSION

Detailed measurements confirm the impression that axons
do not grow randomly, even in homogeneous environments, and
the average fractal dimension of $D = 1.28-1.33$ shows that
axons tend to grow straight (Fig. 3 and TABLE I). The determination of a specific objective number—the fractal dimension—that characterizes the relative straightness of growth
of an axon now permits a quantitative assessment of the effect of various experimental perturbations (e.g., the use of
cell-motility disrupting drugs such as taxol and cytochalasin) on this intrinsic tendency for straight growth (Katz
and George, 1985).

One critical experimental variable is the substrate adhesivity. Substrate adhesivity affects the rate at which axons elongate (Luduena, 1973; Letourneau, 1975a; Collins,
1980), and <u>regional differences</u> in substrate adhesivity can
determine the final patterns of axons (Weiss, 1945; Letourneau, 1975a; Collins and Garrett, Jr., 1980; Akers et al.,
1981; Lander et al., 1982; Rogers et al., 1983; Katz and Lasek, 1985ab). On the other hand, comparisons of elongation
on polyornithine-coated and on uncoated acid-rinsed glass
suggest that <u>uniform</u> changes in substrate adhesivity may, in
themselves, have only a minor effect on the actual straightness of growth (TABLE I).

Although growth cones tend to actively alternate sides,
axons do not grow in convoluted or ''wiggly'' paths. Growth
cones changed their directions of movement significantly
more often than would be expected by pure chance, and the
angles through which they moved spanned the entire range of
a full 180° (Fig. 4). This broad searching behavior of the
growth cone appears to be strongly tempered by the axon
proper, because the axon itself tended to maintain its same
orientation (to continue pointing in the same direction)
throughout its growth (Fig. 5).

One possible cause for the orienting and the damping
effects that the axon proper has on its route of growth is an
intrinsic resistance to bending: the average axonal bend was
only 17°, and, 95% of the time, axons were bent less than 65°
from straight lines. Axons behave somewhat like elastic rods
(Katz, unpublished observations), and the longitudinally-
oriented axonal cytoskeleton may be the prime determinant of
these rod-like properties of the axon (Gilbert, 1975; Morris
and Lasek, 1982 1984). In embryonic axons, the bulk of the
core cytoskeleton is formed of microtubules (Bray and Bunge,
1981; Letourneau, 1982), and microtubules have an inherent
elastic stiffness (Kamimura and Takahashi, 1981). The longi-

tudinal rigidity of the axonal cytoskeleton may be the major cause of the inherent tendency of axons to grow straight.

The intrinsic tendency for axons to grow straight confirms the notion that when axons grow in convoluted patterns ex-trinsic factors must be the major determinants of the final axonal configuration. As Harrison (1910, pp. 837/835) wrote:

The outgrowth of the nerve fibers . . . can be accounted for by power of growth in a straight line modified by deflection as a result of minor obstacles in the path. . . . Given a form of protoplasm with power to extend itself in a definite direc-tion so as to form a fiber, the next step is to determine the influences which may modify the direction of its growth and produce the specific arrangement of nerve tracts found in the mature organism.

At the other end of the spectrum, certain regions of the nervous system (such as the early circumferential fibers of the spinal cord (Holley, 1983)) are characterized by sheets of parallel straight axons (Katz and Lasek, 1985ab). In these regions, the intrinsic tendency of axons to grow straight may be one of the critical determinants of the fi-nal axon patterns.

ACKNOWLEDGMENTS

I am indebted to W. F. Eddy for many constructive con-versations about the problems of analyzing the straightness of growth of axons, to T. Hoshiko for reminding me about fractals, to R. J. Lasek for first raising the question of how straight axons actually grow, and to L. J. Gilbert and L. F. Watson for excellent technical assistance. This work was supported by grants from the National Institutes of Health and the Whitehall Foundation and by an Alfred P. Sloan Foundation Research Fellowship.

112   SAMPLE PAPER

REFERENCES

Akers, R. M., Mosher, D. F., and Lilien, J. E. (1981) Promotion of retinal neurite outgrowth by substratum–bound fibronectin. Dev. Biol. 86: 179–188.

Bray, D. (1973) Branching patterns of individual sympathetic neurons in culture. J. Cell Biol. 56: 702–712.

Bray, D. (1979) Mechanical tension produced by nerve cells in tissue culture. J. Cell Sci. 37: 391–410.

Bray, D. and Bunge, M. B. (1981) Serial analysis of microtubules in cultured rat sensory axons. J. Neurocytol. 10: 589–605.

Collins, F. (1980) Neurite outgrowth induced by the substrate associated material from nonneuronal cells. Dev. Biol. 79: 247–252.

Collins, F. and Garrett, Jr., J. E. (1980) Elongating nerve fibers are guided by a pathway of material released from embryonic nonneuronal cells. Proc. Natl. Acad. Sci. (USA) 77: 6226–6228.

Gilbert, D. S. (1975) Axoplasm architecture and physical properties as seen in the Myxicola giant axon. J. Physiol. 253: 257–301.

Gundersen, R. W. and Barrett, J. N. (1980) Characterization of the turning response of dorsal root neurites toward nerve growth factor. J. Cell Biol. 87: 546–554.

Gundersen, R. W. and Park, K. H. C. (1984) The effects of conditioned media on spinal neurites: substrate–associated changes in neurite direction and adherence. Dev. Biol. 104: 18–27.

Hamburger, V. (1960) A Manual of Experimental Embryology, rev. ed., Univ. of Chicago Press, Chicago.

Harrison, R. G. (1910) The outgrowth of the nerve fiber as a mode of protoplasmic movement. J. Exp. Zool. 9: 787–848.

Heacock, A. M. and Agranoff, B. W. (1977) Clockwise growth of neurites from retinal explants. Science 198: 64–66.

Holley, J. A. (1983) Development of the circumferential axonal pathway. PhD Thesis. Case Western Reserve Univ., Cleveland, OH.

Kamimura, S. and Takahashi, K. (1981) Direct measurement of the force of microtubule sliding in flagella. Nature 293: 566–568.

Katz, M. J. (1984) Stereotyped and variable growth of redirected Mauthner axons. Dev. Biol. 104: 199–209.

Katz, M. J. and George, E. B. (1985) Fractals and the analysis of growth paths. Bull. Math. Biol., in press.

Katz, M. J., George, E. B., and Gilbert, L. J. (1984) Axonal elongation as a stochastic walk, Cell Motil. 4: 351–470.

Katz, M. J. and Lasek, R. J. (1979) Substrate pathways which guide growing axons in Xenopus embryos. J. Comp. Neurol. 183: 817–832.

Katz, M. J. and Lasek, R. J. (1985a) How are the elemental axon patterns produced in the spinal cord? In Perspectives of Neuroscience: From Molecule to Mind, Tsukada, Y., ed., Univ. Tokyo Press, Tokyo, pp. 43–60.

Katz, M. J. and Lasek, R. J. (1985b) Early axon patterns of the spinal cord: experiments with a computer. Dev. Biol. 109: 140–149.

Katz, M. J. and Watson, L. F. (1984) Intensifier for bodian stain of tissue sections and cell cultures. Stain Technol., 60: 81–87.

Lander, A. D., Fujii, D. K., Gospodarowicz, D., and Reichardt, L. F. (1982) Characterization of a factor that promotes neurite outgrowth: evidence linking activity to a heparin sulfate proteoglycan. J. Cell Biol. 94: 574–585.

Lasek, R. J., Metuzals, J., and Kaiserman-Abramof, I. R.

(1983) Cytoskeletons reconstituted in vitro indicate
that neurofilaments contribute to the helical structure
of axons. In Developing And Regenerating Vertebrate
Nervous Systems, Markwald, R. R. and Kenny, A. D.,
eds., Alan R. Liss, NY, pp. 1–18.

Letourneau, P. C. (1975a) Possible roles for cell–to–substratum adhesion in neuronal morphogenesis. Dev. Biol.
44: 77–91.

Letourneau, P. C. (1975b) Cell–to–substratum adhesion and
guidance of axonal elongation. Dev. Biol. 44: 92–101.

Letourneau, P. C. (1982) Analysis of microtubule number and
length in cytoskeletons of cultured chick sensory neurons. J. Neurosci. 2: 806–814.

Luduena, M. A. (1973) Nerve cell differentiation in vitro.
Dev. Biol. 33: 268–284.

Mandelbrot, B. B. (1977) Fractals: Form, Chance, and Dimension. W. H. Freeman, NY.

Mandelbrot, B. B. (1983) The Fractal Geometry of Nature.
W. H. Freeman, NY.

Morris, J. R. and Lasek, R. J. (1982) Stable polymers of the
axonal cytoskeleton: the axoplasmic ghost. J. Cell
Biol. 92: 192–198.

Morris, J. R. and Lasek, R. J. (1984) Monomer–polymer equi-
libria in the axon: direct measurement of tubulin and
actin as polymer and monomer in axoplasm. J. Cell Biol.
98: 2064–2076.

Nieuwkoop, P. D. and Faber, J. (1967) Normal Table of Xenopus laevis (Daudin), 2d ed., North Holland Publ. Co.,
Amsterdam.

Rakic, P. (1971) Neuron–glia relationship during granule
cell migration in developing cerebellar cortex. A Golgi
and electronmicroscopic study in Macacus rhesus. J.
Comp. Neurol. 141: 283–312.

Ramon y Cajal, S. (1911) Histologie du systeme nerveux de

Publication Manual. Washington, DC: American Psychological Association.

Style Manual for Guidance in the Preparation of Papers Published by the American Institute of Physics. New York: American Institute of Physics.

JOURNAL ABBREVIATIONS FOR REFERENCE LISTS

Chemical Abstracts Service Source Index. Washington, DC: American Chemical Society.

List of Journals Indexed in Index Medicus. Bethesda, MD: National Library of Medicine.

Serial Sources for the Biosis Data Base. Philadelphia, PA: Bio-Sciences Information Services.

SCIENTIFIC FIGURES

Papp, C. S. (1976). *A Manual of Scientific Illustration*. American Visual Aid Books, Box 28718, Sacramento, CA 95828.

Tufte, E. R. (1983). *The Visual Display of Quantitative Information*. Graphics Press, Box 430, Cheshire, CT 06410.

STATISTICS

General

Box, G. E. P., Hunter, W. G., and Hunter, J. S. (1978). *Statistics for Experimenters: An Introduction to Design, Data Analysis & Model Building*. New York: John Wiley.

Devore, J. (1982). *Probability & Statistics for Engineering and the Physical Sciences*. Monterey, CA: Brooks-Cole.

Freedman, D., and Pisani, R. (1978). *Statistics*. New York: Norton.

Guttman, I., Wilks, S. S., and Hunter, J. S. (1982). *Introductory Engineering Statistics*, 3rd ed. New York: John Wiley.

Mosteller, F., Fienberg, S. E., and Rourke, R. E. K. (1983). *Beginning Statistics with Data Analysis*. Reading, MA: Addison-Wesley.

Savage, I. R. (1968). *Statistics: Uncertainty and Behavior*. Boston: Houghton Mifflin.

Specific Topics

Jenkins, G. M., and Watts, D. G. (1968). *Spectral Analysis & Its Application*. Oakland, CA: Holden-Day.

Lehmann, E. L. (1975). *Nonparametrics: Statistical Methods Based on Ranks*. Oakland, CA: Holden-Day.

Kennedy, W., and Gentle, J., eds. (1980). *Statistical Computing*. New York: Dekker.

Ripley, B. D. (1981). *Spatial Statistics*. New York: John Wiley.

Tukey, J. W. (1977). *Exploratory Data Analysis*. Reading, MA: Addison-Wesley.

Weisberg, S. (1980). *Applied Linear Regression*. New York: John Wiley.

Scientific Tables

Diem, K., and Lentner, C., eds. (1970). *Scientific Tables*. Ardsley, NY: Geigy Pharmaceuticals.

Weast, R. C., ed. *CRC Handbook of Chemistry and Physics*. Boca Raton, FL: CRC Press.

APPENDIX C

SOME STANDARD SCIENTIFIC ABBREVIATIONS

acceleration due to gravity (9.8 m/sec^2)	g
adenosine 3',5'-cyclic monophosphate	cyclic AMP
adenosine 5'-triphosphate	ATP
alternating current	ac
ampere	A
angstrom	Å
approximately	approx.
atmosphere	atm
boiling point	b.p.
calorie	cal
centigrade (celsius)	°C
centimeter	cm
circular dichroism	c.d.
compare	cf.
complementary DNA	cDNA
counts per minute	cpm
curie	Ci
cycles per second	Hz
deoxyribonucleic acid	DNA
direct current	dc
disintegrations per minute	dpm
ethylenediaminetetraacetic acid	EDTA
gram	gm
half-life	$t_{1/2}$
high-pressure liquid chromatography	HPLC
hour	hr
hydrogen ion concentration ($-\log [H^+]$)	pH

if and only if	iff
immunoglobulin	Ig
intramuscular	i.m.
intraperitoneal	i.p.
intravenous	i.v.
joule	J
kilogram	kg
kilometer	km
limit	lim
liter	liter
logarithm, base 10	log
logarithm, base e	ln
maximum	max.
melting point	m.p.
meter	m
messenger RNA	mRNA
microgram	mg
microliter	μl
micrometer	μm
milliampere	mA
milligram	mg
milliliter	ml
millimeter	mm
minimum	min.
minute	min
molar (moles/liter)	M
mole	mole
molecular weight	mol. wt.
month	mo
nanometer	nm
normal concentration	N
nuclear magnetic resonance	NMR
number	no.
page	p.
pages	pp.
parts per million	ppm

percent	%
picometer	pm
polyacrylamide gel electrophoresis	PAGE
radioimmunoassay	RIA
revolutions per minute	rpm
ribonucleic acid	RNA
ribosomal RNA	rRNA
second	sec
species (singular)	sp.
species (plural)	spp.
specific gravity	sp.gr.
standard deviation	SD
temperature (in equations)	T
thin-layer chromatography	TLC
time (in equations)	t
transfer RNA	tRNA
ultraviolet	UV
volt	V
volume	vol.
watt	W
week	wk
weight	wt.
year	yr

Notes:
• All abbreviations (except p./pp., sp./spp.) are the same for singular and plural units, e.g., the abbreviation for *years* is *yr*, not *yrs*.
• Use abbreviations for units only when preceded by a number; write: "4 yr," but "many years."

APPENDIX D

GREEK ALPHABET

Name	Symbol	
	Lower case	Upper case
alpha	α	A
beta	β	B
gamma	γ	Γ
delta	δ	Δ
epsilon	ε	E
zeta	ζ	Z
eta	η	H
theta	θ	Θ
iota	ι	I
kappa	κ	K
lambda	λ	Λ
mu	μ	M
nu	ν	N
xi	ξ	Ξ
omicron	o	O
pi	π	Π
rho	ρ	P
sigma	σ	Σ
tau	τ	T
upsilon	υ	Y
phi	φ	Φ
chi	χ	X
psi	ψ	Ψ
omega	ω	Ω

APPENDIX E

power of 10	prefix	symbol
10^9	giga	G
10^6	mega	M
10^3	kilo	k
10^{-1}	deci	d
10^{-2}	centi	c
10^{-3}	milli	m
10^{-6}	micro	μ
10^{-9}	nano	n
10^{-12}	pico	p

MATHEMATICAL CONSTANTS

$$\text{pi} = 3.14159 \qquad e = 2.71828$$

ELEMENTARY MATHEMATICAL FORMULAE

Logarithms:

$$\log xy = \log x + \log y$$

$$\log x^n = n \log x$$

$$\log_a x = (\log_b x)/(\log_b a)$$

Trigonometry: $\sin A = a/c$

$\cos A = b/c$

$\tan A = a/b$

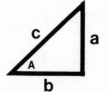

Algebra:

$$\sum_{i}^{n} i = n(n+1)/2$$

binomial coefficient $C = n!/(k!(n-k)!)$

(C is the number of combinations of n things taken k at a time, in any order)

APPENDIX F

CRITICAL t VALUES

(see Appendix B for sources of more complete tables)

$N-1$*	0.9†	0.5	0.2	0.1	0.05	0.01
1	0.16	1.00	3.08	6.31	12.71	63.66
2	0.14	0.82	1.87	2.92	4.30	9.93
3	0.14	0.77	1.64	2.35	3.18	5.84
4	0.13	0.74	1.53	2.13	2.78	4.60
5	0.13	0.73	1.48	2.02	2.57	4.03
6	0.13	0.72	1.44	1.94	2.45	3.71
7	0.13	0.71	1.42	1.90	2.37	3.50
8	0.13	0.71	1.40	1.86	2.31	3.36
9	0.13	0.70	1.38	1.83	2.62	3.25
10	0.13	0.70	1.37	1.81	2.23	3.17
12	0.13	0.70	1.36	1.78	2.18	3.06
14	0.13	0.69	1.35	1.76	2.15	2.98
16	0.13	0.69	1.34	1.75	2.12	2.92
18	0.13	0.69	1.33	1.73	2.10	2.88
20	0.13	0.69	1.33	1.73	2.09	2.85
25	0.13	0.68	1.32	1.71	2.06	2.97
30	0.13	0.68	1.31	1.70	2.04	2.75
60	0.13	0.68	1.30	1.67	2.00	2.66
120	0.13	0.68	1.29	1.66	1.98	2.62
∞	0.13	0.67	1.28	1.65	1.96	2.58

*degrees of freedom
†P values (significance levels)

APPENDIX G

CRITICAL X^2 VALUES

(see Appendix B for sources of more complete tables)

$N-1$*	0.9†	0.5	0.2	0.1	0.05	0.01
1	0.02	0.46	1.64	2.71	3.84	6.64
2	0.21	1.39	3.22	4.61	5.99	9.21
3	0.58	2.37	4.64	6.25	7.81	11.34
4	1.06	2.75	5.99	7.78	9.49	13.28
5	1.61	4.35	7.29	9.24	11.07	15.09
6	2.20	5.35	8.56	10.64	12.59	16.81
7	2.83	6.35	9.80	12.02	14.08	18.48
8	3.49	7.34	11.03	13.36	15.51	20.09
9	4.17	8.34	12.24	14.68	16.92	21.67
10	4.87	9.34	13.44	15.99	18.31	23.21
12	6.30	11.34	15.81	18.55	21.03	26.22
14	7.79	13.34	18.15	21.06	23.69	26.12
16	9.31	15.34	20.47	23.54	26.30	32.00
18	10.87	17.34	22.76	25.99	28.87	34.81
20	12.44	19.34	25.04	28.41	31.41	37.57
25	16.47	24.34	30.68	34.38	37.65	44.31
30	20.60	29.34	36.25	40.26	43.77	50.89
40	29.05	39.34	47.27	51.81	55.76	63.69
50	37.69	49.34	58.16	63.17	67.51	76.15
75	59.80	74.33	85.07	91.06	96.22	106.39
100	82.36	99.33	111.67	118.50	124.34	135.81
150	128.28	149.33	164.35	172.58	179.58	193.21
200	174.84	199.33	216.61	226.02	233.99	249.45

*degrees of freedom
†P values (significance levels)

APPENDIX H

Random Number Table

```
1 8 4 4 4   2 4 5 1 3   5 6 9 5 0   2 1 4 8 9   0 7 8 6 2
5 2 6 6 1   5 3 2 1 6   2 4 1 2 5   9 4 8 1 4   1 7 6 2 8

6 5 3 8 0   1 8 3 0 6   3 5 1 3 6   3 9 5 4 6   2 7 7 6 4
0 9 0 8 4   5 2 1 1 3   5 2 0 3 9   2 7 8 0 7   4 5 6 3 6

5 8 7 4 1   5 6 1 8 8   6 5 3 4 8   5 7 2 9 1   1 3 9 6 1
2 7 6 4 5   8 9 0 2 7   4 1 5 3 5   1 7 8 8 9   6 5 1 4 0

3 9 1 1 8   8 4 0 1 8   6 0 1 0 4   7 7 6 2 1   3 1 4 1 5
1 5 3 9 8   6 1 7 2 8   4 2 0 5 3   9 8 4 3 6   6 5 9 3 7

9 4 6 8 5   4 9 3 1 1   6 0 2 5 1   3 7 0 8 4   8 0 3 1 7
4 8 3 1 3   4 2 8 8 6   1 0 7 2 9   0 6 1 3 4   6 6 5 1 0

1 2 4 4 6   1 6 5 9 3   5 6 4 7 6   2 2 5 3 1   3 0 4 1 3
7 0 6 6 6   4 0 6 8 2   1 9 6 0 8   4 6 1 4 4   7 6 4 2 1

5 5 0 5 7   2 0 7 2 5   2 1 5 9 4   2 3 0 1 8   7 7 0 0 8
2 8 0 2 9   7 3 5 5 1   9 8 9 5 8   3 3 8 1 5   0 1 6 1 9

9 1 8 4 1   7 8 0 6 5   4 4 9 1 3   2 7 5 4 9   1 0 8 4 4
1 0 8 2 4   6 4 3 1 7   9 6 2 2 1   6 3 8 2 7   2 9 4 7 8

5 8 3 1 1   1 2 8 6 2   4 2 2 0 6   4 5 4 3 5   4 9 8 9 0
8 9 4 7 1   6 8 0 7 1   3 9 0 0 3   5 5 8 6 0   0 7 9 1 6

6 7 5 8 9   9 3 1 7 0   8 0 7 7 0   5 3 8 3 2   7 4 1 9 7
8 3 4 2 8   0 1 4 1 4   3 7 9 0 0   7 4 8 9 5   4 6 5 0 7

1 9 4 0 5   7 8 0 6 4   6 8 8 1 6   6 5 3 9 1   3 6 9 1 4
0 2 5 0 6   7 7 5 9 4   3 6 5 8 0   4 4 4 8 1   6 1 1 1 7

5 3 3 4 7   5 4 0 7 1   1 7 9 4 7   5 8 3 1 8   2 8 1 2 2
0 9 6 0 4   0 1 2 8 9   0 6 2 6 0   2 7 3 8 4   0 4 7 7 6

9 8 7 7 2   9 4 1 3 2   2 0 3 2 9   2 6 8 4 6   9 1 9 3 0
2 1 8 9 8   5 8 1 7 2   1 2 9 4 0   0 6 6 8 6   5 5 3 4 4

4 0 2 7 1   2 4 7 1 2   3 9 8 4 9   2 3 2 2 9   8 7 4 5 5
9 2 7 3 1   3 4 0 0 7   0 8 5 8 9   1 6 1 9 7   7 4 8 4 4

0 3 5 5 0   0 3 1 4 1   0 4 3 4 2   5 9 3 8 6   7 0 0 2 4
4 0 3 0 2   6 7 7 7 4   9 0 3 7 4   7 6 9 9 8   7 4 6 4 6

6 0 7 5 6   5 9 5 4 7   8 8 0 9 9   4 7 7 4 2   0 6 0 8 7
8 7 7 0 4   7 7 0 9 2   0 4 1 1 0   9 0 3 1 5   7 9 4 7 9

8 9 1 2 8   8 4 6 0 7   0 6 3 7 4   6 6 4 4 3   5 6 9 6 5
4 1 7 4 0   6 0 5 1 1   2 9 3 5 3   4 7 1 3 9   7 4 7 3 8

2 1 7 3 7   9 6 0 3 6   7 9 6 2 6   6 1 1 0 2   8 9 1 8 8
5 0 0 6 0   4 5 9 8 1   3 6 4 1 5   8 8 7 8 9   9 4 8 7 3

5 1 4 7 1   2 3 7 2 4   8 4 4 5 5   8 8 3 0 9   6 0 0 0 2
1 8 5 4 2   6 4 3 2 0   7 9 3 0 6   9 0 7 7 9   8 3 9 9 8
```

APPENDIX I

Common Nontechnical Science Words*

active (activity)
analysis (analyze, analytical)
appear (appearance)
compare (comparison)
contain
control
data
determine
differ (different, differences) / similar (similarity)
effect
effective
evidence
experiment
identify (identification)
important (importance)
increase / decrease
indicate
lack
level
observe (observation)
occur
produce
product (production)
rate
regulate (regulation, regulatory)
result (resultant)
sequence (sequential)
show
study
suggest (suggestive)
type

*Most common nontechnical science words found in eight science papers published in 1984 in the *Journal of the American Medical Association, Nature, Proceedings of the National Academy of Sciences (USA)*, and *Science* and covering the disciplines of cell and molecular biology, ecology, epidemiology, geology, medicine, meteorology, and physiology.

Index